Practical Universe

Observations, Experiments, Exercises

Second Edition

Manfred Cuntz
Nilakshi Veerabathina
Levent Gurdemir
James Davis

Department of Physics
The University of Texas at Arlington

Kendall Hunt
publishing company

Kendall Hunt
publishing company

Contents

Preface

Eight years have passed since the first edition of *Practical Universe: Observations, Experiments, Exercises*. Not unexpectedly, there have been dramatic developments in Astronomy and Astrophysics in the meantime. They include research aimed at the Solar System (particularly Mars), stars and galaxies, as well as the Universe at large. Most recently, the first Earth-sized planet outside of the Solar System (Kepler-186f) was discovered and gravitational waves were identified by the LIGO observatory; the latter further enhances our physical understanding of Black Holes. Currently, intense research efforts are focused on dark matter and dark energy, which appear to be decisive influences on the large-scale structure and evolution of space. Admittedly, most of these results are of primary interest to professionals and thus outside of the scope of introductory astronomy labs – but nonetheless, quite often, they serve as inspiration to students and instructors alike to become more deeply involved in Astronomy.

Compared to the previous edition of *Practical Universe*, all units have been revised and rewritten. This is a reflection of our teaching experience and the feedback from our students and lab assistants; the latter have provided tireless efforts during the course of the recent years. Since the first edition of *Practical Universe*, more than 5000 students have signed up for Introductory Astronomy I and II at The University of Texas at Arlington. As science and teaching styles continue to develop, we added several new units to the book, and retired a few others instead. New units include *Astrophotography*, the identification of *Moon and Mars Landing Sites,* and the calculation of *Habitable Zones* for different star-planet systems. The latter unit deals with estimating the potential of habitability for possible Earth-type planets in those systems, considering that habitability is assumed to require the existence of liquid water on the planetary surface. In total, three of the book's units now focus on exoplanets.

In the future, we expect Astronomy to continue to develop even further. The authors of *Practical Universe* are looking forward to future teaching endeavors, knowing that there will be exiting times ahead. Comments from our students will be always warmly welcome – just like in the past.

UNIT 1: MATHEMATICAL TOOLS

UNIT 1.1 POWERS OF TEN

OBJECTIVE

To learn to represent ordinary numbers in scientific notation, perform simple calculations using powers of ten, and stating numbers with proper significance

INTRODUCTION

The science of astronomy encompasses both the truly gargantuan – billion light year long strings of galaxies, black holes hundreds of millions of times heavier than the sun, and stars larger than the orbits of some planets – as well as the microscopic, like the masses of atoms and the sizes of tiny dust grains that formed the planets. For example, the largest black hole has a calculated mass of nearly 2,000,000,000,000,000,000,000,000,000,000,000,000 kilograms. The hydrogen atoms which make up stars have tiny masses, only 0.00000000000000000000000000167 kilograms per atom. Trying to express these numbers in standard long form requires unnecessary space and invites minor writing errors that could drastically influence calculations and produce very wrong results. Scientists use a far more efficient and standardized way of representing very large or very small numbers by consolidating all of the leading or trailing zeroes in a compact form called scientific notation.

Powers of Ten

Scientific notation takes advantage of the powers of ten and is based on two fundamental rules:

1. A positive exponent denotes the number of zeroes following the 1. For example, 10^2 is a 1 followed by two zeroes, making this number 100. 10^6 is a 1 followed by six zeroes, which represents one million. 10^0 is a one followed by no additional zeroes, representing the number 1.
2. A negative exponent is used for numbers smaller than one and denotes the number of zeroes falling to the right of the decimal point, including the 1. For example, 10^{-2} is scientific notation for 0.01; 10^{-4} is 0.0001.

Since the powers of ten notation omits all the cumbersome zeros, a wide range of numbers can be compactly represented, as shown in the following tables (from 10^{-12} to 10^{12}). The tables include the scientific notation, the long form of that number, the spoken translation of the number, that number's commonly used prefix (such as kilo for kilometer, meaning 1000 meters), and an example of something in the physical world which would be measured by a measurement of that magnitude.

Negative Powers

Power	Value	Name	Prefix	Example
10^{-1}	0.1	Tenth	deci	Seconds required to circle the Earth at light speed
10^{-2}	0.01	Hundredth	centi	Seconds required to cross the US at light speed
10^{-3}	0.001	Thousandth	milli	Mass of Jupiter compared to the mass of the Sun
10^{-4}	0.0001	Ten Thousandth		Mass of the largest asteroid compared to Earth's mass
10^{-5}	0.00001	Hundred Thousandth		Distance from the Earth to the Sun, in light years
10^{-6}	0.000001	Millionth	micro	Wavelength of red light, in meters
10^{-7}	0.0000001	Ten Millionth		Area of a computer screen pixel, in square meters
10^{-8}	0.00000001	Hundred Millionth		Width of a DNA strand, in meters
10^{-9}	0.000000001	Billionth	nano	Odds of winning the Powerball lotto
10^{-10}	0.0000000001	Ten Billionth		Width of a water molecule, in meters
10^{-11}	0.00000000001	Hundred Billionth		Speed of continental drift, in meters per second
10^{-12}	0.000000000001	Trillionth	pico	Mass of a human cell, in kilograms

Zero and Positive Powers

Power	Value	Name	Prefix	Example
10^0	1	One		Height of the average human being, in meters
10^1	10	Ten	deca	Height of a giraffe, in meters
10^2	100	Hundred	hecto	Height of a skyscraper, in meters
10^3	1,000	Thousand	kilo	Mass of a car, in kilograms
10^4	10,000	Ten Thousand		Mass of a 1 cubic meter slab of gold, in kilograms
10^5	100,000	Hundred Thousand		Size of the Milky Way Galaxy, in light years
10^6	1,000,000	Million	mega	Weight of the Saturn V moon rocket, in kilograms
10^7	10,000,000	Ten Million		Number of seconds in one year
10^8	100,000,000	Hundred Million		Distance from Earth to the asteroid belt, in kilometers
10^9	1,000,000,000	Billion	giga	Current population of China or India
10^{10}	10,000,000,000	Ten Billion		Age of the Universe, in years
10^{11}	100,000,000,000	Hundred Billion		Number of stars in the Milky Way Galaxy
10^{12}	1,000,000,000,000	Trillion	tera	Distance to the next closest star, in kilometers

Note: In British English and German, a billion is 10^{12} and a trillion is 10^{15}.

Conversion of a Number to Scientific Notation

The process for converting a number from long form to ordinary scientific notation requires a simple, standard two-step process:

1. Move the decimal point to the position that is immediately after the *first non-zero* digit (if no decimal point is expressly written, it is assumed to be at the end of the number)
2. Determine the exponent (power of ten) by counting the number of places the decimal point moves. The power increases (positive) if the decimal point moves to the *left* and decreases (negative) if the decimal point moves to the *right*

Examples:

$$167{,}000 \rightarrow \frac{\text{decimal needs to move}}{\text{5 places to the left}} \rightarrow 1.67 \times 10^5$$

Note: while identical in terms of value, 0.167×10^6 and 16.7×10^4 and $167. \times 10^3$ are not written in proper standard scientific notation, which requires the decimal to fall after the first non-zero digit.

$$0.00335 \rightarrow \frac{\text{decimal needs to move}}{\text{3 places to the right}} \rightarrow 3.35 \times 10^{-3}$$

In these cases, the scientific notation contains two parts: the coefficient term (1.67 and 3.35) and the magnitude term (the power of ten term, given as 10^5 and 10^{-3}, respectively).

Mathematical Operations with Scientific Notation

Some lab exercises may require calculations with scientific notation. Mathematical operations performed using very large or very small numbers written in ordinary notation are tedious and prone to calculator input errors. Scientific notation makes calculation on these numbers simpler and more straightforward. For the example of math using scientific notation, assume that a and b represent ordinary numbers.

- To multiply powers of 10, add the exponents: $10^a \times 10^b = 10^{(a+b)}$
  Example: $10^{92} \times 10^{11} = 10^{(92+11)} = 10^{103}$
- To divide powers of 10, subtract the exponents: $\dfrac{10^a}{10^b} = 10^{(a-b)}$
  Example: $\dfrac{10^4}{10^7} = 10^{(4-7)} = 10^{-3}$
- To raise powers of 10 to other powers, multiply exponents: $\left(10^a\right)^b = 10^{(a \times b)}$
  Example: $\left(10^9\right)^3 = 10^{(9 \times 3)} = 10^{27}$

Multiplications and Divisions

When performing multiplication or division on powers of ten with coefficients, treat the coefficient and magnitude terms separately. For multiplication, multiply the coefficients and add the exponents.

Example: Multiply 600 and 50,000

Translate these in proper scientific notation: $\left(6 \times 10^2\right) \times \left(5 \times 10^4\right)$

Multiply coefficients and add exponents: $[6 \times 5] \times [10^{2+4}] \rightarrow [30] \times [10^6]$

Rewrite the answer into standard scientific notation: $30 \times 10^6 \rightarrow \boxed{3 \times 10^7}$

10^7 is translated as ten million, so this number is read as 30 million. In long form, this math would have required inputting 600×50000 and obtaining 30000000 on a calculator. The process is equally straightforward for division. Separate coefficient from magnitude, divide the coefficients, and subtract the exponents:

Example: Divide 360,000 by 200

Translate the numbers into scientific notation, split coefficients from magnitudes, perform the division operations, and obtain the final answer.

$$\frac{3.6 \times 10^5}{2 \times 10^2} \rightarrow \left[\frac{3.6}{2}\right] \times \left[\frac{10^5}{10^2}\right] \rightarrow [1.8] \times [10^{5-2}] \rightarrow 1.8 \times 10^3$$

Additions and Subtractions

When adding and subtracting scientific notation, rewrite the numbers so that the coefficients have the same magnitude, adjust the coefficient by moving the decimal right or left, as necessary. Factor out the power of ten (which is now the same for both numbers), add the coefficients, and then rewrite the number in proper scientific notation.

Example: Add the Earth's mass in kilograms (5.972×10^{24}) to the mass of Jupiter in kilograms (1.898×10^{27})

$$\left(5.972 \times 10^{24}\right) + \left(1.898 \times 10^{27}\right)$$

$$\left(5.972 \times 10^{24}\right) + \left(1898 \times 10^{24}\right)$$

$$\left(5.972 + 1898\right) \times \left(10^{24}\right) \rightarrow \left(1903.972\right) \times \left(10^{24}\right) \rightarrow \boxed{1.904 \times 10^{27}}$$

In this example, both numbers were adjusted to matching powers of 10^{24}; adjusting the coefficients so that both coefficients have magnitudes of 10^{27} would also yield the same answer

$$\left(0.005972 \times 10^{27}\right) + \left(1.898 \times 10^{27}\right)$$

$$\left(1.898 + 0.005972\right) \times \left(10^{27}\right) \rightarrow \left(1.903972\right) \times \left(10^{27}\right) \rightarrow \boxed{1.904 \times 10^{27}}$$

The process is the same even if one magnitude is positive and one is negative:

Example: Add 9.3×10^1 and 7.7×10^{-1}

$$\left(9.3 \times 10^1\right) + \left(7.7 \times 10^{-1}\right)$$

$$\left(9.3 \times 10^1\right) + \left(0.077 \times 10^1\right)$$

$$\left(9.3 + 0.077\right) \times \left(10^1\right) \rightarrow \left(9.377\right) \times \left(10^1\right) \rightarrow \boxed{9.377 \times 10^1}$$

Significant Figures

An important aspect to any scientific field of study is the idea of expressing uncertainty in the results. Any observation, measurement, or reading will include some uncertainty, as no device or laboratory set-up is capable of exact measurement with infinite precision. On very large or small scales, irregularities in the object being studied or the device doing the measuring will lead to uncertainties in values. Every physical measurement involves some approximation at some level, and significant figures express how confident scientists are in their quoted measurements.

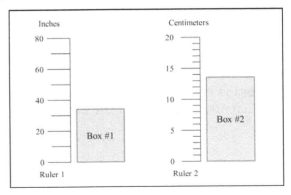

As a rule of thumb, the error in reading laboratory equipment is ±0.1 or 1/10 of the smallest division marked on the measuring instrument. For example, consider the schematic of two boxes being measured with two separate rulers:

Ruler 1 uses inches, part of the Imperial (non-metric) measurement system. The spacing between consecutive visible tick marks measures 10 inches, so the uncertainly is ±1 inch (10% of the 10 inches spacing). Therefore, one person may approximate the height of the box as 35 inches, while another may read it as 34 inches. Both answers would be correct to within the measurement error of ±1 inch.

Ruler 2 utilizes the metric system of centimeters. The spacing of the tick marks is significantly finer, allowing for a more precise measurement with a smaller uncertainty. The smallest marked unit on the ruler is 1 centimeter, leading to an error of ±0.1 centimeter (0.1 centimeters = 1 millimeter in the metric system). Box #2's upper end falls somewhere between 13 and 14 centimeters. The true height of Box #2 falls somewhere within the range of 13.5±0.1 centimeters.

The idea of "error" in a measurement also puts a cap on how precisely a number can be measured. For example, given that the height of Box #2 can only be estimated down to ±0.1 centimeters, it isn't possible to give an estimate below this threshold.

Everyone who uses the ruler to measure Box #2 would agree that the height is more than 13 centimeters but less than 14 centimeters. Therefore, anyone asked to measure the box would always include the numbers 1 and 3 in their answer. Deviation would come into play in the decimal point, however. Some may estimate 13.4, or 13.5, or 13.6 centimeters. Because of the limitations of the ruler, the third digit would vary slightly from reader to reader. That third digit would be an estimate or an approximation, open to some doubt about its accuracy. It is thus called the doubtful digit. Doubtful does not mean incorrect, but means that the estimate lies at the ruler's threshold for precision. Scientists would have a slightly lower confidence level in that digit but would include it as a significant (measurable and reliable) part of the measurement. For this reason, scientists would say this number has 3 significant figures.

There is no possibility that someone could accurately measure a fourth digit. An answer of 13.48 would raise some suspicion. The 1 and 3 would be certain and the 4 would be reasonable. The 8, however, represents 1/100 of a centimeter and is far below the ability of Ruler 2 to accurately measure. The 8 would be a total guess, with no confidence that it is an accurate, measured part of the height. This is the concept of the *significant figure*: it is the part of a measurement in which scientists have a high level of confidence.

Rules for Working with Significant Figures

When reading a measurement and determining significant figures, keep in mind:

1. Non-zero numbers are always significant
2. Leading zeroes are never significant
3. Embedded zeroes are always significant (zeroes between non-zero numbers)
4. Trailing zeroes are only significant if a decimal point is specified

As a note, writing a number in scientific notation makes seeing which figures are significant and which are not significant much easier to determine. The magnitude (the power of ten term) is not counted toward significant figures. When considering significance in a number, consider just the coefficient term.

Addition or subtraction: the last digit retained in an answer is set by the first doubtful

Multiplication or division: the answer contains no more significant figures than the *least* accurately known number.

The table below contains detailed examples of significant figures with an explanation of why the digits are or are not considered significant.

Number	Number of Significant Figures	Scientific Notation	Explanation
0.0088	2	8.8×10^{-3}	Leading zeroes are not significant
5.055	4	5.055×10^{0}	Embedded zeroes are significant
600	1	6×10^{2}	Trailing zeroes without a specifically included decimal are not significant
600.	3	6.00×10^{2}	Trailing zeros here are significant because of the explicitly specified decimal point
600.0	4	6.000×10^{2}	Trailing zeros here are significant because of the explicitly specified decimal point
0.00600	3	6.00×10^{-2}	Leading zeroes are never significant; trailing zeros here are significant because of the specified decimal point

Calculating Answers with Significant Figures

If no measurement can ever have infinite precision, the same can be said for calculations which use these measurements. When performing calculations, answer should be given with proper number of significant figures, to express how dependable – or how approximate – your answer actually is. This table below will demonstrate how to determine the significant figures to be included in your answer.

Addition

Final answers can only contain 1 doubtful number. Complete the addition and writing your final answer from left to right, stop at the first doubtful number. In the sum below, each measurement has its doubtful digit highlighted in bold:

$$
\begin{array}{r}
3.123\mathbf{4} \\
+ \quad 1.01\mathbf{1} \\
+ \quad 5.3\mathbf{8} \\
\hline
9.5144
\end{array} = \boxed{9.51}
$$

3.1234 has its doubtful digit in the ten-thousandths place (0.0004); 1.011 has its doubtful digit in the thousandths place; 5.38 has its doubtful digit in the hundredths place. Writing the answer from left to right, stop at the first doubtful digit; this in this case is the digit 1 in the hundredths place. The final answer ends up with 3 significant figures (with the last number being doubtful).

Subtraction

The same process applies as with addition. The final answer may only contain one doubtful digit.

$$
\begin{array}{r}
2.378\mathbf{9} \\
- \quad 1.9845\mathbf{7} \\
\hline
0.3943\mathbf{3}
\end{array} = \boxed{0.3943}
$$

In this example, the first doubtful digit occurs at the ten-thousandths place, so all digits up to that point can be kept. The final answer ends up with 4 significant figures (with the last number being doubtful).

Multiplication

When multiplying two numbers, the final answer must be given with the same number of significant figures as your least accurately known number. In the same way that a chain is only as strong as its weakest link, a calculated value is only as precise as the least precise measurement.

$$5.3322 \times 4.2 = 22.39524 \rightarrow \boxed{22}$$

While one number contains 5 significant figures, the second number only contains 2 significant figures. Therefore, the final answer has only 2 significant figures.

Division

When dividing two numbers, the same rule of least significance applies. The significant figures in your answer must match the number of significant figures in your least precise measurement.

$$\frac{8.677244}{1.9} = 4.5669705 \rightarrow \boxed{4.6}$$

Again, despite the high precision of one number – with 7 significant figures – the final answer still has low precision due to the 2 significant figures of the second measurement. Notice also that the first doubtful digit in the final answer has been rounded up, from 5 to 6, giving a final answer of 4.6; always round up if the first number after the doubtful digit is greater than or equal to 5.

Name _____ Id _____

Due Date _____ Lab Instructor _____ Section _____

Worksheet # 1

In the Significant Figures column, record the number of significant figures in the given number. In the Scientific Notation column, write down the number in proper scientific notation, making sure to use the proper number of significant figures.

	Number	Significant Figures	Scientific Notation
1.	347		
2.	206265		
3.	365.256366		
4.	0.000678		
5.	3.14159		
6.	149.6 million		
7.	6500 billionths		
8.	4.55 billion		
9.	320.00×10^4		
10.	6		
11.	22.3×10^7		
12.	0.075 trillion		
13.	19.81×10^{-6}		
14.	0.08×10^{-3}		
15.	Fifty four billion		

Name _____ Id _____

Due Date _____ Lab Instructor _____ Section _____

Worksheet # 2

Write the following numbers in scientific notation, keeping the specified number of significant figures and rounding up where necessary.

1. Express the number 99.995 with 2 significant figures

2. Express the number 0.00004920575 with 4 significant figures

3. Express the mass of the Sun (1,988,950,000,000,000,000,000,000,000,000 kg) three ways: with 5 significant figures, 3 significant figures, and 2 significant figures.

Complete the math below, giving your final answer in standard scientific notation with proper significant figures.

1. $\left(9.13353 \times 10^2\right) \times \left(1.1 \times 10^1\right)$

2. $\left(4.00 \times 10^{-2}\right) \times \left(5.3 \times 10^{-1}\right)$

3. $\left(6.03 \times 10^{56}\right) \times \left(2.2809336 \times 10^{77}\right)$

4. $\dfrac{7.322 \times 10^4}{8.85 \times 10^2}$

5. $\dfrac{7.322 \times 10^4}{8.85 \times 10^{-2}}$

6. $\dfrac{10000}{553.00727}$

7. $12.655 + 0.91942$

8. $1.820 - 1.81118$

Name _____ Id _____

Due Date _____   Lab Instructor _____   Section _____

Worksheet # 3

1. At a museum, the skeleton of a *Tyrannosaurus rex* is on display. A visitor reads the plaque and learns that radiometric dating places the age of the bones at 65,000,000 years old. When the visitor returns 12 years later, he is confused as to why the plaque has not been updated to indicate that the skeleton is now 65,000,012 years old. He assumes that the museum curator is just too lazy to update the information. Considering significant figures, carefully and clearly state why the visitor is either correct or incorrect in his assertion that the skeleton is now 65,000,012 years old.

2. In a laboratory, a scientist makes measurements and finds a sample to have a length of 6 centimeters and a width of 2 centimeters. What is the surface area of the sample, written with proper significant figures, given that the surface area is equal to *length* × *width*?

UNIT 1.2   MEASURING ANGLES

OBJECTIVE

To become familiar with angles, their measurements, conversion, and use in astronomy

INTRODUCTION

In geometry, any two connected points form a line. Two connected lines create an angle between them. Before the advent of telescopes (which allowed for detailed study of objects too small to see with the naked eye) astronomy as a science was limited to drawing star charts and carefully measuring the positions of planets against the background stars. Even with telescopes, careful studies of positions and measured angles allowed astronomers to determine the sizes of the planets and the distances between orbits.

Angles are measured in the unit of degrees. One full circle, rotation, or revolution contains 360 degrees. In much the same way as an hour is composed of smaller units called minutes, 1 degree is composed of smaller units called *arcminutes*. One arcminute is an incredibly small angle. Two lines which deviate by 1 arcminute would appear so close to one another that they be mistaken for one line to your naked eye. In fact, only after drawing the lines nearly 300 inches (25 feet) long would you see the ends of the lines separated by 1 inch, making an extremely long sliver of a triangle with a 1 inch long side.

Just as 1 minute of time is further broken down into 60 individual seconds, 1 arcminute is composed of 60 equally spaced subunits called *arcseconds*. Two lines which make an angle of 1 arcsecond to each other would have to be drawn out incredibly long before their deviation was noticeable. At 3.25 miles long, the ends of the two lines would lie one inch apart.

Angular diameter and angular separation are two critical concepts in astronomy and they both utilize the measurement of angles. The angular separation describes how distant two objects *appear* from one another in the sky. From our vantage point on Earth, the sky appears flat and two dimensional. The distance we see between stars is an angular separation, not an actual three dimensional, linear distance. For example, the three stars of Orion's belt appear very close to one another, making a straight line in the sky. In actuality, those three stars are each hundreds of light years away from the Earth and even further apart from one another. Their relative closeness (small angular separation) is an optical illusion. Along the same vein, when planets "align" during a conjunction and appear extremely close to one another in the sky, they are actually tens of millions of miles apart.

Angular diameters describe what percentage of your vision is occupied by an object. Human vision spans about 180° across. Something incredibly small – like the head of a pin from 100 yards away – takes up so little of your vision as to be invisible. The head of a pin viewed from a football field away would cover about 1 arcsecond of vision. The Great Wall of China – seen from orbit – is only about 20 arcminutes wide, still too small for the receptors in the eye to recognize. Your fist – held out at arm's length – will cover roughly 10 degrees of your vision. In this way, angular diameter describes how large something appears to be rather than its absolute size.

PROCEDURE

Measuring the angular separation between two lines requires a protractor. The standard protractor is a semi-circle marked off with ticks running from 0° to 180°. To measure the angle made by two lines, place the center of the protractor on the intersection point of the two lines.

Adjust the protractor until one of the lines is pointing directly at the angle of 0°. The second line will point to the reading representing the angle between the lines. It may make measurements easier to use a ruler and extend the length of the lines (which make the angle) so that you do not have to estimate where the measured line points.

Likewise, to draw an angle, begin by drawing two points and connecting them with a straight line. This will serve as the baseline. Place the center of the protractor at one end of the line, making sure that the line points to 0°. Mark a third point on your protractor at the desired angular separation in degrees and use the straight edge of a ruler to connect those points. Smaller units are one arcmin (1′) and one arcsec (1″).

EQUATIONS AND CONSTANTS

$$1° = 60'$$

$$1° = 3600''$$

$$1' = \frac{1}{60}°$$

$$1'' = \frac{1}{60}' = \frac{1}{3600}°$$

Name _____ Id _____

Due Date _____   Lab Instructor _____   Section _____

Worksheet # 1

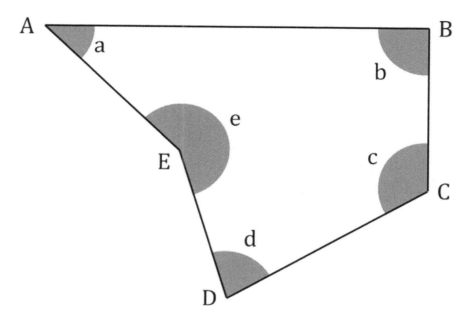

Below is a 5 sided geometric shape made up of lines AB, BC, CD, DE, and EA. At the point where the lines meet they make angles a, b, c, d, and e, with the angle highlighted in gray. Use a protractor to measure the indicated angles and give an answer in degrees.

1. Angle **a** (the angle between lines AB and AE)

2. Angle **b** (the angle between lines AB and BC)

3. Angle **c**

4. Angle **d**

5. Angle **e**

Name _____ Id _____

Due Date _____   Lab Instructor _____   Section _____

Worksheet # 2

Using a protractor, draw a pair of lines intersecting at the listed angle.

1.   17°

2.   68°

3.   91°

4.   166°

5.   345°

Name _____ Id _____

Due Date _____ Lab Instructor _____ Section _____

Worksheet # 3

Use the conversions to determine the following angles.

1. In the Earth's sky, the moon has an angular diameter of 0.5°. What is the angular diameter of the Moon in arcminutes?

2. Through a telescope, two stars are separated by 2° 15′ 35″. What is their separation in arcseconds?

3. The stars Mizar and Alcor – both in the handle of the Big Dipper – are located 12′ 21″ from one another. What is that separation in arcseconds?

4. Arrange these angular diameters in order of largest to smallest angular diameter:
    a) 0.25°
    b) 90′
    c) 1000″

Name _____ Id _____

Due Date _____   Lab Instructor _____ Section _____

Worksheet # 4

The four dashed circles below represent the orbits of the innermost planets, with S representing the position of the sun, V the position of Venus, E the position of Earth, and M the position of Mars at some given time.

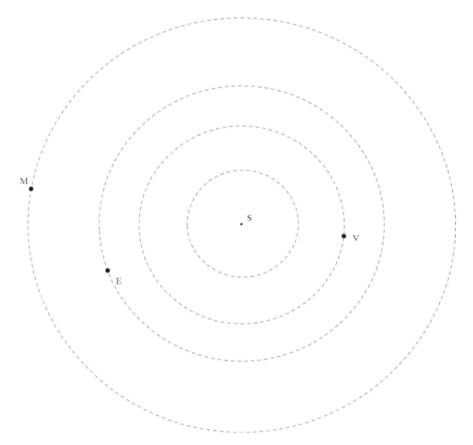

1. From the perspective of someone on Earth, how far apart would the Sun and Mars appear, in degrees? (What is the angle from M to E to S?) How far apart would the Sun and Venus appear, in degrees? (What is the angle from S to E to V?)

2. Someone on Earth determines that Mercury is 16° from the Sun. Mark Mercury's possible position in orbit with an *M* (Mercury's orbit is the innermost orbit.)

UNIT 1.3   TEMPERATURE SCALES

OBJECTIVE
To become acquainted with the relations between the three most common temperature scales: Celsius, Fahrenheit and Kelvin

INTRODUCTION
Temperature is a measure of the energy of the particles which make up a body or an environment, like a cup of water or the atmosphere of a star, or rock on the surface of a planet. The temperature is related to the average speed of atoms or molecules which make up an object. A hot object, like the water molecules in a cloud of steam, are moving rapidly and chaotically, with each molecule containing relatively large amounts of energy. By contrast, the low-velocity atoms of water ice are held stationary in a solid crystal structure. As the ice is heated, the velocity of the molecules begins to break apart those bonds, reducing the crystal structure to a liquid puddle and eventually an expanding gas cloud.

Three temperature scales are most commonly used in everyday life, science, and industry.

The **degree Celsius (°C)** temperature scale was devised by the Swedish astronomer Anders Celsius in 1742. This scale is based on the behavior of pure water, which freezes at 0 °C and boils at 100 °C under standard atmospheric conditions (at sea level on the Earth). Therefore, there are 100 degrees between these points. This scale is used throughout the science and almost in all countries of the world except the United States.

The **Kelvin (K)** temperature scale, named after the British Lord Kelvin (William Thomson, Baron Kelvin of Largs) is an extension of the degree Celsius scale down to *absolute zero or 0 K,* a hypothetical temperature at which all atomic and molecular motions cease. *Absolute zero* is the lowest temperature hypothetically possible at which no heat exists. On the Kelvin temperature scale, water freezes at 273 K (more precise value: 273.16 K) and boils at 373 K. Absolute zero (0 K) or –273 °C is the starting point for the Kelvin scale. Since nothing can be colder than 0 K, there are no negative temperatures on the Kelvin scale.

The step size of the Kelvin and the Celsius temperature scales is same, because water must be heated by 100 K or 100 °C to go from its freezing to melting point. Scientists throughout the world (including the United States) prefer the Kelvin scale because it is closely related to the physical meaning of the temperature.

Note: Temperatures on this scale are called "kelvin," *not* degrees kelvin. Further, kelvin is *not* capitalized, and the symbol (capital K) stands alone with no degree symbol.

The **degree Fahrenheit (°F)** temperature scale, now antiquated, is still used by many in the United States. The German physicist Gabriel Fahrenheit introduced this scale in the early 1700s and he intended 0 °F to represent the coldest temperature achievable at that time and 100 °F to represent the temperature of a healthy human body. As you might know, normal body temperature is closer to 98.6 °F, suggesting that when he conducted his experiment, either he was having a fever, or his thermometer was inaccurate. Lastly, it is believed that he might have used a cow's temperature instead of his own. On this scale, water freezes at 32 °F and boils at 212 °F. Therefore, there are 180 degrees between these points. The step size of the Fahrenheit degree is smaller than that of the Celsius degree (or 1 kelvin). In other words, a degree Celsius (or a kelvin) is 180/100 (which is 9/5 or 1.8 times) the size of the degree Fahrenheit. An increase of 1 kelvin is equivalent to a Fahrenheit temperature increase of nearly 2 degrees. Notice that the degree Fahrenheit is a non-metric temperature scale, while the degree Celsius and the Kelvin temperature scales are metric scales (based on multiples of 10).

Note that the United States is the only country that uses Fahrenheit temperatures for shelter-level (surface) weather observations. However, since July 1996 all surface temperature observations in the National Weather Service METAR/TAF reports are transmitted in degrees Celsius.

EQUATIONS AND CONSTANTS

Fahrenheit to Celsius Conversion

$$C = \frac{(F - 32)}{1.8}$$

Fahrenheit to Kelvin Conversion

$$K = \frac{(F - 32)}{1.8} + 273.16$$

Celsius to Fahrenheit Conversion

$$F = (C \times 1.8) + 32$$

Celsius to Kelvin Conversion

$$K = C + 273.16$$

Kelvin to Fahrenheit Conversion

$$F = (K - 273.16) \times 1.8 + 32$$

Kelvin to Celsius Conversion

$$C = K - 273.16$$

Name _____ Id _____

Due Date _____ Lab Instructor _____ Section _____

Worksheet # 1

Answer the following questions related to the temperature scale

1. Which temperature scale or scales begin at zero?

2. Which temperature scale or scales allow for negative temperatures?

3. At what temperature does water freeze:
   On the Fahrenheit scale:

   On the Celsius scale:

   On the Kelvin scale:

4. A 1 degree temperature change on the Fahrenheit scale is equal to how many degree of temperature change on the Celsius scale?

5. Normal human body temperature is 98.6 °F; what is the healthy human body temperature:
   On the Celsius scale:

   On the Kelvin scale:

6. The color of a hot metal is directly related to the temperature of the metal. The coil on a stovetop burner will begin to glow a dim, deep red at 390 °C. At what temperature does a stove coil begin to glow:
   On the Celsius scale:

   On the Kelvin scale:

Continue....

7. The boiling point of oxygen – where oxygen transitions from a liquid puddle into a gas cloud – occurs at –183 °C. At what temperature does that process happen:

    On the Fahrenheit scale:

    On the Kelvin scale:

8. When temperatures drop below 63.15 K, nitrogen freezes, turning from liquid nitrogen into solid ice. At temperatures above 77.36 K, liquid nitrogen turns into a gas. On the dwarf planet Pluto, the wintertime temperature is low enough to freeze the nitrogen atmosphere, turning the atmosphere into a gentle snow of ice crystals. In the summertime, the temperature is high enough to turn nitrogen into a gas. If a thermometer on the surface of Pluto reads –214.15 °C, show whether Pluto's nitrogen will be frozen into ice, existing in puddles, or in a gaseous form making up an atmosphere.

9. The Sun's surface temperature is 5770 K. What is the temperature of the Sun:

    In Fahrenheit:

    In Celsius:

10. Absolute 0 on the Kelvin scale is 0 K. What is this in Fahrenheit?

11. At the beginning of the week, the temperature is measured at a chilly 40 °F. At the end of the week, the temperature has risen to 80 °F. Has the temperature doubled? Explain why of why not. (As a hint, consider look at question 10 and consider where the Fahrenheit scale begins).

UNIT 1.4   MASS, WEIGHT, AND DENSITY

OBJECTIVE
To understand the make-up of objects from measurements of mass and density

INTRODUCTION
In everyday language, people use the words "mass" and "weight" interchangeably without considering that there is a significant difference between the two.

Mass (m) corresponds to the total amount of matter (or total number of atoms) in an object. Just like time and length, mass is a fundamental quantity that does not change with changing location, position, movement, or shape. Mass may only be altered by the addition or subtraction of material. The International Standard (SI) metric unit for mass is the kilogram (kg), with 1 kg = 1000 grams.

Weight (W) is the measure of force of attraction that is acting on an object. In the case of Earth, this force is caused by the acceleration due to gravity (or simply, gravity). The effect of gravity (g) on mass (m) describes the weight (W) of an object ($W = mg$).

Density (ρ) describes the amount of matter contained in a given volume. The density of a type of material under normal conditions is a constant. The density is a function of how compacted the matter in a sample is and the mass and size of the atoms making up that material (hence, gold, made of very heavy atoms, is a very dense material; carbon, with a small atomic nucleus made of six protons and six neutrons, is a lightweight material). In a liquid or gaseous form, dense materials will sink to the center of mass, while lower density material will buoy to the surface. It is thought that the fact that ice is less dense than water allowed life to flourish in Earth's ancient tide pools and shallow bodies of water, since freezing water floats, insulating the water from losing heat. If ice was denser than water, frozen lake water would sink to the bottom, crushing anything in its way (including fledging life) and – with no insulation from cold air – more layers of water would lose heat energy and freeze.

Using the density of a planet as a starting point, astronomers can make educated predictions as to the composition of the unseen interior of a planet. For example, tiny Mercury – the smallest of the planets – is extremely dense, nearly twice as dense as common terrestrial rock. Therefore, astronomers could postulate that Mercury's interior is a large, high density sphere of iron covered by a thin layer of solid rock. The same can been seen with Jupiter's moon Europa, which is substantially less dense than rock and is actually covered by a deep, deep liquid water ocean covered by a thick layer of solid ice. Using this technique, astronomers have an important tool in understanding the make-up of planets, comets, moons, and asteroids.

The Earth and other terrestrial planets are composed primarily of light, rocky material (carbon, silicon, aluminum) and dense metallic material (iron, nickel, cobalt). Scientists, though, are unable to burrow into the Earth to confirm its composition (and the same can certainly be said of the other planets) yet they are able to make general statements about the interior composition of the major planets. Among other aspects, this lab will explore the concepts of density and mass for simple, ordinary materials.

EQUATIONS AND CONSTANTS

$$1 \text{ cm} = 10 \text{ mm}$$

$$1 \text{ kg} = 1000 \text{ g}$$

$$1 \text{ kg/m}^3 = 1000 \text{ g/cm}^3$$

Density

$$\rho = \frac{m}{V}$$

Volume of a Sphere (*r*: radius)

$$V = \frac{4}{3}\pi r^2$$

Volume of a Rectangular Solid (*l*: length, *w*: width, *h*: height)

$$V = l \cdot w \cdot h$$

Volume of a Cylinder or Disk (*r*: radius, *h*: height)

$$V = \pi r^2 h$$

DATASHEET #1

Below are the densities of some common substances

Substance	Density (kg/m³)	Density (g/cm³)
Liquid Hydrogen	68	0.068
Gasoline	730	0.730
Ice	917	0.917
Pure Water	1000	1.000
Seawater	1025	1.025
Aluminum	2700	2.700
Granite	3000	3.000
Iron	7870	7.870
Lead	11340	11.34
Gold	19300	19.30

Name _____ Id _____

Due Date _____   Lab Instructor _____   Section _____

Worksheet # 1

For each of the following questions, make sure to include all applicable units and show your work.

1. If you have cubes of 1 cm³ consisting of aluminum, iron, and lead, which of the following will be true (check all that apply)?
   a) All three cubes will have equal density
   b) The aluminum cube would be the densest
   c) The aluminum cube would have the smallest volume
   d) The lead cube would have highest mass
   e) All three cubes would have the same mass
   f) If melted into a liquid and poured together, aluminum would float to the top
   g) If melted into a liquid and poured together, iron would sink to the bottom

2. A cube of tungsten is 2.0 centimeter wide by 2.0 centimeters long by 2.0 centimeters high, and a balance determines that the mass is 154 grams.

   What is the volume of the cube?

   Calculate the density of tungsten:

3. The cube (see question 2), is cut directly in half along its length and the excess material is discarded.

   What are the new dimensions of the cube?

   What is the mass of this sample?

   What is the weight of this sample?

   Calculate the density of tungsten:

Continue....

4. The sample of tungsten (see question 2) is altered again. It is cut into three equal strips, and two of the strips are discarded. The remaining strip's material is remolded into a sphere. That sphere is cut into quarters. What is the density now?

5. The Earth is a sphere with a radius of 6,378,000 meters and a mass of 5.974×10^{24} kilograms.

   What is the density of the Earth?

   Looking at the chart of common material, which of those matches Earth's density (within 10% of the density you calculated)?

   If an object was 50% iron and 50% rock (granite), what would the density of that object be?

   What can you tentatively conclude about the Earth's structure and composition based on your previous answer?

6. The planet Saturn is nearly circular, with a radius of 120,536,000 meters and a mass of 5.685×10^{26} kilograms.

   How many times more massive is Saturn than the Earth?

   What is the density of Saturn?

   Why do some textbooks say that Saturn would float if placed in a large enough pool of water?

Name _____ Id _____

Due Date _____ Lab Instructor _____ Section _____

Worksheet # 2

PROCEDURE:

Check that you have the following equipment: balance; metal samples; rock samples; Vernier scale; large and small graduated cylinder

For part 1, use the Vernier scale to carefully measure the dimensions of the metal samples and record those in the table provided. The metal samples are precisely machined, with easily measured lengths, heights, and diameters. Measurements and calculations should use units of centimeters and grams

For part 2, use the balance to determine the mass of the rock samples. Since the rocks are not simple or regular geometric shapes, you may not calculate their volume by direct measurement. Instead, fill the graduated cylinders as close to halfway with water as possible. Gently release the first rock sample into the cylinder note the new water level. Every 1 mL of displacement is equal to 1 cm³ of volume. Remove the rock, reset the water level, and continue with the second sample.

Part 1: Density of Metals

Material	Length	Height	Width	Mass	Volume	Density

Part 2: Density of Rocky Materials

Material	Mass	Volume	Density

Continue....

1. What was your most dense metallic material? What was the least dense? How many times more dense was the denser material?

2. Earth's density is given as 5.51 g/cm³ and outwardly appears to be composed of rock. How does that density compare to the densities of your rocks? How much more dense is the Earth than your rock samples' average density?

3. How does Earth's density compare with iron? How much more dense is iron compared to Earth?

4. If you took the average density of your rocks and iron, what would that be? How does that compare to Earth's density? What does that say about the Earth's structure and composition?

UNIT 1.5   ASTRONAUTS' WEIGHTS

OBJECTIVE

To consider the gravitational strength on different planets and to calculate the weight of astronauts on those different worlds

INTRODUCTION

The weight of an astronaut on a planet (including our home planet Earth) is determined by the gravitational acceleration on the surface of that planet, g. The gravitational surface acceleration itself is given by the mass and the size of the planet. The sizes of planets can be expressed by planetary diameters or planetary radii. (As studied in the previous Unit 1.4, the mass and size of a planet also determine the planet's density.) Clearly, the concept of gravitational strength does not just apply to different planets (including planets outside of the Solar System), but also to moons, including Earth's Moon. For example, the gravitational strength on the Moon is considerably smaller compared to Earth. Astronauts training on Earth were burdened down with hundreds of pounds of equipment, making walking and moving difficult. However, on the Moon, those same astronauts were able to handle massive equipment with ease.

PROCEDURE

Please consider: the weight of an astronaut on a general planet w_{planet} is given by his/her weight on Earth w_{Earth} by taking into account the gravitational accelerations on that planet and on Earth, g_{planet} and g_{Earth}, respectively, as shown below.

EQUATIONS AND CONSTANTS

$$1 \text{ m} = 1000 \text{ mm}$$

$$1 \text{ kg} = 1000 \text{ g}$$

$$G = 6.67259 \times 10^{-11} \text{ m}^3/\text{kg·s}^2$$

Acceleration

$$a = \frac{v}{t} \quad \text{or} \quad a = \frac{2x}{t^2}$$

Gravitational Acceleration

$$g = \frac{GM}{r^2}$$

Weight

$$w_{planet} = w_{Earth} \frac{g_{planet}}{g_{Earth}}$$

Name _____ Id _____

Due Date _____ Lab Instructor _____ Section _____

Worksheet # 1

This worksheet deals with the practical way in which scientists measured the gravitational strength of the Earth (i.e., the gravitational acceleration) and the mass of the Earth through the use of simple experiments and observations.

Determining Earth's gravitational acceleration by experiment: As the story goes, Galileo Galilei wanted to disprove the idea that heavier objects fell faster than lighter objects. While this looks true (a dropped sheet of paper will drift to the ground while a boulder with come crashing down) Galileo realized very light objects are held up by motions of the air (called air resistance). He could show that a 10 kilogram weight and 50 kilogram weight – dropped from the same height – will hit the ground simultaneously. So he scrambled to different loges (open balcony levels) of the Tower of Pisa, measured the height from the ground, and carefully timed the number of seconds between release and impact for the weights.

Dropped From:	Height (m)	Fall Time (s)	Acceleration (m/s²)
Loggia 1	10.4	1.5	
Loggia 2	16.2	1.8	
Loggia 3	22.2	2.1	
Loggia 4	27.0	2.4	
Loggia 5	33.6	2.6	
Loggia 6	39.1	2.8	
Loggia 7	45.4	3.0	
Roof	46.5	3.1	
Belfry	54.6	3.3	

Average acceleration from the data points	

1. Determine Earth's mass from gravitational acceleration: As far back as the days of ancient Greece, astronomers were able to calculate the diameter and radius of the Earth (such as by measuring the length of shadows at different latitudes to geometrically infer the curvature and size of the planet). In Isaac Newton's time, the Earth's radius was calculated as 6,370 km with its gravitational acceleration calculated to be 9.8067 m/s². Using these two values and Newton's Gravitational Acceleration equation, calculate the mass of the Earth in kg.

Name _____ Id _____

Due Date _____ Lab Instructor _____ Section _____

Worksheet # 2

Here we consider the case of Apollo astronauts training on Earth prior to their visit to the Moon. On Earth, they were exposed to heavy loads of equipment, making walking and moving difficult, a situation entirely different when walking on the Moon.

1. Using the Moon's radius of 1,736,500 meters and lunar mass of 7.348×10^{22} kg, determine the Moon's gravitational acceleration (g)

2. Determine the weight that each of the following would have had on the lunar surface:

Apollo Astronauts: Earth vs. Moon

Object	Earth Weight (lbs)	Lunar Weight (lbs)
Astronaut's Own Body	170	
Space Suit	250	
5 lb Rock Hammer	5	
Lunar Buggy	470	
Lunar Seismometer	130	
Lunar Lander	33,000	

3. Determine the gravitational acceleration (g) and weight of the astronauts (170 lbs on Earth) on other planets, using the planetary masses and radii given in the attached tables.

   (The Sun is included for the sake of comparison only.)

Planet	Radius = Diameter / 2 (m)	g (m/s²)	g ratio (g_{planet}/g_{Earth})	Astronaut's Weight (lbs)
Venus				
Mars				
Jupiter				
The Sun				

UNIT 2: UNAIDED EYE OBSERVATIONS

UNIT 2.1 INTRODUCTION TO THE NIGHT SKY

OBJECTIVE
To learn to use astronomical software to identify objects in the night sky

INTRODUCTION
Once the sun has set and the sky is no longer flooded with the sun's scattered light, the stars and planets of the night sky become apparent. As the Earth rotates, some objects will set on the western horizon and others will rise in the east. The presence of streaking artificial satellites and meteors make the sky even more dynamic. For many, however, city lights blot out most of the sky with "light pollution," making seeing the disk of the Milky Way Galaxy, faint meteors, and even moderately bright stars impossible. Even without leaving the city limits and seeking out darker skies, sky charts, planetarium software, and internet sources can give observers an idea of what is in the night sky. Some useful sources are listed below:

Internet Sources:

- http://www.space.com/skywatching/
- http://www.skyandtelescope.com/observing/sky-at-a-glance/
- http://hubblesite.org/explore_astronomy/tonights_sky/

NASA's virtual telescope:

- http://skyview.gsfc.nasa.gov

Computer Software:

- Stellarium: (www.stellarium.org)
- Celestia: (www.shatters.net/celestia/)
- Digital Universe Atlas: (http://www.amnh.org/our-research/hayden-planetarium/digital-universe)

Software like Stellarium or the Digital Universe Atlas can be used to simulate the passage of time, track celestial objects, and view events such as eclipses, meteor showers, and planetary alignments. This lab will walk through the use of Stellarium's user controls to track planets through the night sky. The program itself was written for use in urban and suburban schools where students didn't have easy access to dark skies or a planetarium, and was built up and adapted to include more information about the night sky, such as the temperature of a particular star or the distance to a targeted galaxy.

PROCEDURE

Once Stellarium has been downloaded and installed, open the program. Keyboard commands can be used to alter the view of the night sky. For example, pressing the **A** key will toggle the atmosphere on and off. Pressing the **7** key will bring time to a stop. Bringing up the help window (either by clicking on the question mark icon on the left-hand tool bar or pressing the **F1** key) will list all of the possible keyboard commands for Stellarium. **F7** will also bring up the long list of keyboard commands.

By default, Stellarium will choose your computer's clock time as the starting time and date and Paris, France, as the starting location.

Pressing **F6** brings up the location window, allowing you to set your location either from a list of world cities or from latitudes and longitudes.

Pressing **F5** brings up the date/time window, allowing you to set a desired date and time. From this window, you can also adjust the time by increasing or decreasing individual years, months, days, hours, minutes, or seconds

Other useful commands include:

- **A** turns the atmosphere on and off, removing the obscuring effect of sunlight
- **B** turns on the official boundaries of the modern day constellations
- **C** turns on the constellation guide lines, generally used to map out the rough shape of the constellation traced out the bright stars
- **G** turns the ground on and off
- **H** turns on the horizon line, showing the exact position of the transition between the sky and ground
- **7** stops the flow of time; **K** sets the flow of time to normal
- **V** turns on the labels and names of the constellations
- **F11** switches between full screen and small window views of the program
- **PageUp** and **PageDown** increases and decreases the magnification (zooms in and out on the center of the screen)
- **= [the equal sign key]** jumps time forward by 24 hours
- **, [the comma key]** turns on the ecliptic line, the path of the sun through the sky
- **\ [the backslash key]** snaps to normal (non-zoomed) field of view
- **F3** brings up the object search window; pressing **Ctrl** and **F** at the same time has the same effect

Stop the flow of time and set and record the starting date and location. Turn on the ecliptic line. All of the major planets, along with the sun and moon, will fall in close proximity to this line. Use the mouse or arrow keys to sweep your vision through the sky. Make a note of which planets are already above the horizon and visible on that night.

Only five planets (not including the Earth) are visible to the unaided eye. Choose one of the planets by opening the Search Window and entering the planet's name. The program will center that planet in your field of view, even if it is below the horizon. Determining their exact position in reference to the horizon is made easier by turning the ground animation off and turning the horizon line on. Move time forward or backwards using the date and time window arrows until the planet is just above (for rising) and just below (for setting) the horizon line.

Name _____ Id _____

Due Date _____   Lab Instructor _____   Section _____

Worksheet # 1

Starting Location:	

Starting Date:		Starting Time:	21:00

Which planets are already above the horizon at 9:00 PM?

Mark the location, rise time, and set time of the planets on the chosen date.

Planet	Constellation	Rise Time	Set Time	Apparent Magnitude	Distance from Earth
Mercury					
Venus					
Mars					
Jupiter					
Saturn					

UNIT 2.2   USE OF STAR MAPS

OBJECTIVE

To become familiar with the layout of the night sky and to identify stars and constellations

INTRODUCTION

Up until the invention of the astronomical telescope, astronomy was focused almost exclusively on the mapping of stars and planets. It was noticed early in human history that the stars appeared stable and static, with the same stars tracing out the same familiar patterns year after year, decade after decade, and generation after generation. Accurate star position maps could be used to determine the time of year or a geographic position on Earth's surface. Mythological stories of gods and heroes were attached to these reliable, unchanging, seemingly eternal points of light in the sky. The purpose for the naming was twofold. In dedicating parts of the sky to heroes and gods and vital characters from mythology, sky watchers were honoring the gods upon whom they depended for good fortune and favor as well as immortalizing important deeds, events, and moral lessons in the heavens. Practically, dividing the sky into constellations gave other astronomers a reference frame for measurements and seasonal determinations. For example, the presence of the sun in the constellation Gemini would signal the height of summer and the onset of shorter daylight hours day after day.

The sky – much like the Earth's surface – is broken into vertical and horizontal angular slices. On Earth, we call this coordinate system the latitude and longitude system. A combination of latitude and longitude coordinates can locate any position on the Earth surface (such as NASA's Hawaiian W.M. Keck Observatory, located at 19.86° N latitude, 155.48° W longitude; or the European Southern Observatory in La Silla, Chile, located at −29.16° latitude, 70.44° longitude). In the same way, the Right Ascension and Declination coordinate system can locate any object in the sky, such as the coordinates of the North Star, Polaris (2 hours 31 minutes Right Ascension and +89.25° Declination) or the position of the next closest star to the Earth, Proxima Centauri (14 hours 29 minutes RA and –62° Dec). Just as 0° latitude is the equator of the Earth, 0° Declination represents the celestial equator. 90° on Earth is the North Pole, 90° Declination in the sky is the North Celestial Pole. At any point during the night, an observer can see roughly 12 hours of right ascension from the eastern to western horizon and 90° of declination from the northern to southern horizon.

While declination is measured in the familiar angular measurement units of degrees, Right Ascension is measured in hours, minutes, and seconds, which would appear to imply that Right Ascension measures time, rather than location. The logic behind this choice of unit is the apparent motion of the stars through the night sky. As the Earth turns, stars rise and set and move through the sky. A photograph of the sky would show "star trails" as the stars steadily change position. Each hour of time that passes, a star will appear to have moved 15° through the sky. It is not the star that is actually moving but is the Earth which is turning beneath the sky. Every minute that passes, a star arcs 0.25° through the sky.

Positions in the night sky can be approximated using your hand as a simple measuring device. A fist held at arm's length is approximately 10° across. A thumb held up at arm's length has a width of approximately 1.5°. The eraser on a pencil held out at arm's length is approximately 0.5° across. This rough measurement system can be used to estimate star-to-star distances and can identify dimmer stars and constellations according to their separation from brighter stars. For example, the harder to pick out constellation Lepus the Rabbit lies 16° (or about one-and-a-half fist-widths) below Orion's Belt. The tight star cluster the Pleiades lies four fists – or approximately 40° – from the Belt.

Finally, astronomers measure the relative brightness of stars. Even through large telescopes, stars look like little more than points of varying brightness. On a star chart, a very bright star is indicated by a very large point drawn at the star's position. A very dim star is denoted by a small dot. Astronomers use a system called "magnitudes" to more precisely measure a star's light. In the magnitude system, a very bright star has a small (or even negative) magnitude. A star which is very dim, meanwhile, will have a very large magnitude. In this way, the magnitude scale is "backwards," with a larger magnitude value indicating a lower light output.

PROCEDURE

Worksheets #1 and #2 contain a pair of star charts covering the entire equatorial region of the sky from $-55°$ to $+55°$ Declination and running from 0 hours to 12 hours Right Ascension (Worksheet #1) and 12 hours RA to 24 hours RA (Worksheet #2). The major Right Ascension tick marks are labeled with hours, with each minor tick mark representing 6 minutes. The major Declination tick marks are labeled every $10°$, with the minor tick marks covering $1°$. Smaller units are one arcmin ($1'$) and one arcsec ($1''$).

EQUATIONS AND CONSTANT

$$1° = 60'$$

$$1° = 3600''$$

$$1' = \frac{1}{60}°$$

$$1'' = \frac{1}{60}' = \frac{1}{3600}°$$

$$1\ hour = 15°$$

$$1\ min = 0.25°$$

$$Angular\ Separation = \sqrt{RA^2 + Dec^2}_{star\ 1} - \sqrt{RA^2 + Dec^2}_{star\ 2}$$

Name _____ Id _____

Due Date _____   Lab Instructor _____   Section _____

Worksheet # 1

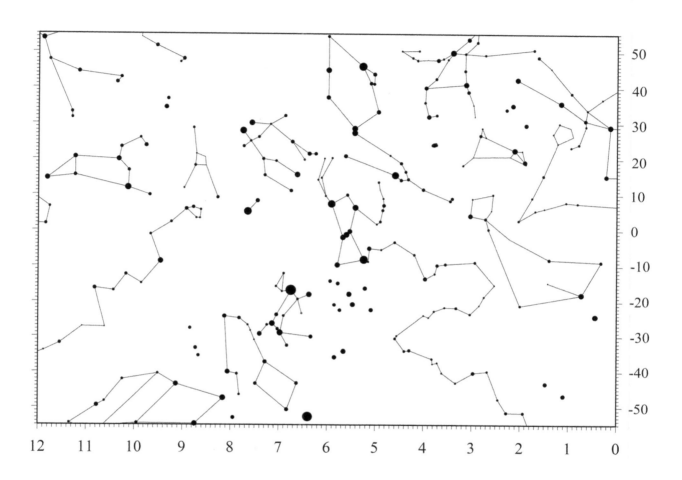

Name _____ Id _____

Due Date _____   Lab Instructor _____ Section _____

Worksheet # 2

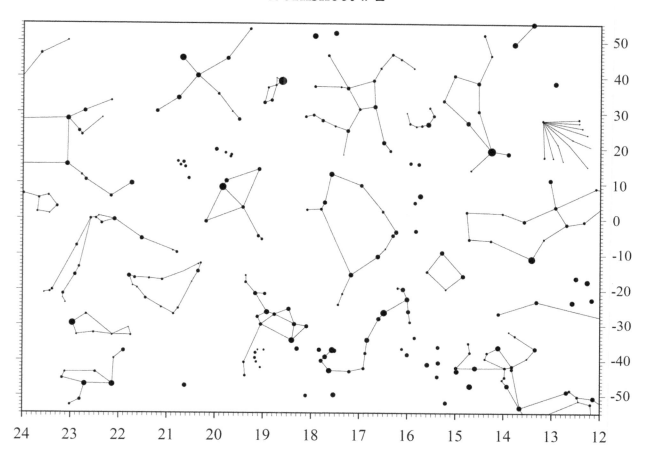

Name _____ Id _____

Due Date _____ Lab Instructor _____ Section _____

Worksheet # 3

Below are listed some of the major constellations visible from the Earth along with the position of their brightest stars and the magnitude (naked eye brightness) of those stars. Using the coordinates, mark each of the stars with their corresponding number and label the constellation with its proper name on Worksheets #1 and #2.

#	Star Name	Constellation	Right Ascension (in hours and minutes)	Declination (in degrees)	Magnitude
1	Sirius	Canis Major	6^h 45^m	$-16°$	-1.45
2	Procyon	Canis Minor	7^h 39^m	$+5°$	0.40
3	Regulus	Leo	10^h 8^m	$+12°$	1.35
4	Pollux	Gemini	7^h 45^m	$+28°$	1.15
5	Antares	Scorpius	16^h 29^m	$-26°$	1.05
6	Betelgeuse	Orion	5^h 55^m	$+7°$	0.45
7	Vega	Lyra	18^h 36^m	$+38°$	0.00
8	Altair	Aquila	19^h 50^m	$+9°$	0.75
9	Spica	Virgo	13^h 25^m	$-11°$	0.95
10	Deneb	Cygnus	20^h 41^m	$+45°$	1.25
11	Arcturus	Bootes	14^h 15^m	$+19°$	0.15

1. Looking at the magnitudes above, which star is the dimmest? Which is the brightest?

2. The constellation Gemini is named after the mythological twins Pollux and Castor. These two stars make up the "heads" of the twins. Using the star chart, estimate the coordinates (RA and Dec) of Castor and record those below.

Continue....

3. The bright red star Betelgeuse is sometimes said to translate into "the armpit" as it appears at the base of the raised left arm of Orion. The equally bright star Rigel is translated as "the foot" of Orion, which is fitting as it appears to be the right foot of the constellation. Using the star chart, estimate the coordinates (RA and Dec) of Rigel and record those below.

4. The Summer Triangle is an easily visible summer sky asterism composed of three of the brightest blue-white stars: Altair, Deneb, and Vega. Connect those stars with a dashed line to mark the Summer Triangle.

5. The Winter Triangle is an easily visible winter sky asterism composed of Procyon, Betelgeuse, and Sirius. Connect those stars with a dashed line to mark the triangle.

Name _____ Id _____

Due Date _____   Lab Instructor _____   Section _____

Worksheet # 4

Deep sky objects are those which are located hundred, thousands, or millions of light years beyond the solar system, and generally require a telescope to spot. However, their location can be generally pointed out in the night sky just by knowing where they fall in relation to brighter nearby stars.

Use a small circle and the indicated letter to the left of the object name, mark each of these objects on the star charts in the location indicated by their coordinates.

	Object	Right Ascension	Declination
A	Andromeda Galaxy	$0^h\ 42^m$	$+41°$
B	Hercules Cluster	$16^h\ 41^m$	$+36°$
C	Cygnus X-1	$19^h\ 58^m$	$+35°$
D	Sagittarius A*	$17^h\ 45^m$	$-29°$
E	51 Pegasi b	$22^h\ 57^m$	$+20°$

THE OBJECTS

Object A is the Andromeda Galaxy, the next closest galaxy to our own Milky Way. At 2.5 million light years distant, it is on a collision course with the Milky Way.

Object B is the Hercules Cluster, a massive, ancient, spherical ball of stars orbiting around the Milky Way galaxy.

Object C is the first object confirmed to be a black hole, with a mass approximately 4.5 million times greater than the Earth's mass.

Object D is the supermassive black hole that sits at the center of the Milky Way galaxy, weighing in at 1.35 trillion times the mass of the Earth.

Object E is the first planet discovered orbiting a star outside of our solar system. About half the mass of Jupiter, 51 Pegasi b orbits its star nearly 95% closer than the Earth orbits the Sun, raising this planet's temperature into the thousands of degrees.

Name _____ Id _____

Due Date _____ Lab Instructor _____ Section _____

Worksheet # 5

The following questions test methods of measuring angles using geometry,

1. The star Betelgeuse has a Right Ascension of 5 hours and 55 minutes. Using the hours-to-degrees conversion, give Betelgeuse's RA in degrees.

2. The star Procyon in Canis Minor has a Right Ascension of 7 hours and 39 minutes. Using the hours-to-degrees conversion, give Betelgeuse's RA in degrees.

3. How many degrees separate Procyon from Betelgeuse in the night sky? (Use the angular separation equation with both RA and Dec in degrees)

4. A good gauge for distances in the night sky is a closed fist held out at arm's length. The width of a fist is approximately 10° across. Approximately how many fists separate Betelgeuse from Sirius, the night sky's brightest star? (You may either calculate this distance using the angular separation equation or by using a ruler to measure the star-to-star distance and comparing that length to the scale on the declination axes of the star chart)

UNIT 2.3   CONSTELLATIONS AND MYTHOLOGY

OBJECTIVE

To learn to identify constellations in the night sky by sight and learn the mythology behind the naming of the constellations

INTRODUCTION

The major recognized constellations of the northern hemisphere are based on characters, stories, and events from Greek mythology. It is a common misconception that the constellations are meant to look like their namesake. In fact, the constellations were named to honor heroes and epic tales of mythology. Therefore, the constellation Perseus was never meant to be viewed as a portrait of the mythological hero; the stars in that specific part of the sky were meant to be dedicated to the memory of Perseus and his amazing feats of heroism. Viewing those distinct, closely spaced stars − Algol, Mirfak, and Atik among others − sky watchers were meant to think about the moral and life lessons taught by the Perseus mythology (which included many characters immortalized in the stars, such as fellow constellations Cassiopeia and Pegasus).

The stars which make up the constellations are generally unrelated to one another, and simply appear close from our vantage point on Earth. Stars which align − like Alnitak, Alnilam, and Mintaka the belt stars of Orion − may actually be vastly different distances from the Earth, or of greatly different ages or luminosities. From different parts of the galaxy, the familiar constellations would be unrecognizable. Since every star in the sky is also orbiting around the Milky Way Galaxy, the passage of thousands of years will eventually cause the stars to drift from their current positions, warping and altering the constellations of which they are a part.

Constellations are more than just the brightest stars that make up the recognizable outlines of the mythological characters. Modern star maps divvy up the sky into 88 distinct sections of various shapes and sizes. All the stars within the borders are considered to belong to that constellation. Orion is more than just the seven stars that outline the vaguely human figure of Orion, but includes 130 stars visible to the naked eye under dark conditions, from the very bright Rigel to the nearly invisible HD 44497. Today, constellations are still used for the naming of stars and identification of regions of space (Alpha Centauri is the brightest star in the direction of the constellation Centaurus; Sagittarius A is the name of the supermassive black hole at the center of the Milky Way Galaxy).

PROCEDURE

In each of the provided tables, record the name of the given constellation along with the meaning or English translation of the constellation's name and the name of the main or brightest star of the constellation. In the space provided, sketch the constellation. This doesn't require great artistic precision, but you should provide a recognizable outline of the constellation. Clearly mark the position of the main star in the constellation.

Name _____ Id _____

Due Date _____ Lab Instructor _____ Section _____

Worksheet # 1

Sketch #1		
	Constellation Name:	
	Translation of Name:	
	Main Star:	

Sketch #2		
	Constellation Name:	
	Translation of Name:	
	Main Star:	

Sketch #3		
	Constellation Name:	
	Translation of Name:	
	Main Star:	

Name _____ Id _____

Due Date _____ Lab Instructor _____ Section _____

Worksheet # 2

Sketch #4		
	Constellation Name:	
	Translation of Name:	
	Main Star:	

Sketch #5		
	Constellation Name:	
	Translation of Name:	
	Main Star:	

Sketch #6		
	Constellation Name:	
	Translation of Name:	
	Main Star:	

Sketch #7		
	Constellation Name:	
	Translation of Name:	
	Main Star:	

UNIT 3: TOOLS FOR ASTRONOMICAL MEASUREMENTS

Unit 3.1 – Lens Optics

Unit 3.2 – Telescopes

Unit 3.3 – Astrophotography

Unit 3.4 – Spectral Analysis

UNIT 3.1   LENS OPTICS

OBJECTIVE

To learn some of the basic aspects of lens-based optical systems like telescopes and binoculars and construct a simple lens telescope

INTRODUCTION

Even though historians agree that Galileo Galilei was not first regarding inventing astronomical telescopes, he made ample use of his first telescope to provide major contributions to science. His first telescope was a simple lens-based system consisting of a tube fixed at both ends by lenses. This primitive telescope was called a refracting telescope. Refraction is the bending of light as it passes through a dense, transparent medium, like water or glass. The forward facing lens was called the *objective*, the rear lens was referred to as the *eyepiece*. The objective had a large surface area capable of intercepting incoming beams of light, bending them, and focusing them into a brighter, concentrated point of light. The spot where the beams of light came together is called the *focal point* (the single point at which the light was focused). The distance from the center of the lens to the spot where the beams of light converge into a sharp, focused point is called the focal length, and is a fixed, physical characteristic of the lens (see illustrations on the next page).

On its own, the objective was not useful for astronomical viewing. The converged light beams would have to be reoriented and directed through the observer's pupil. This was accomplished by the eyepiece lens, a smaller lens that intercepted the light from the objective and redirects it into the observer's pupil.

Magnification is an important aspect of lens-based telescopes intended for casual or amateur observations. Objects like nebula, galaxies, and planets appear small because their light beams only cover a tiny fraction of an observer's field of view. A pair of magnifying lenses gathers the distant object's incoming light and spreads it over a larger swath of the observer's field of view, making it appear larger. The magnification is directly dependent on the focal lengths of the two lenses being used. The larger the focal length of the objective and the smaller the focal length of the eyepiece, the larger the image will appear.

EQUATIONS AND CONSTANTS

Equation	Expression	Variables
Focal Length	$\dfrac{1}{f} = \dfrac{1}{u} + \dfrac{1}{w}$	f : focal length u : distance from light source to lens w : distance from lens to focused image
Focal Length	$f = \dfrac{1}{\dfrac{1}{u} + \dfrac{1}{w}}$	f : focal length u : distance from light source to lens w : distance from lens to focused image
Magnification	$M = \dfrac{f_{obj}}{f_{eye}}$	M : magnification f_{obj} : focal length of objective lens f_{eye} : focal length of eyepiece lens
Minimum Telescope Length	$L = f_{obj} + f_{eye}$	L : distance between eyepiece and objective f_{obj} : focal length of objective lens f_{eye} : focal length of eyepiece lens

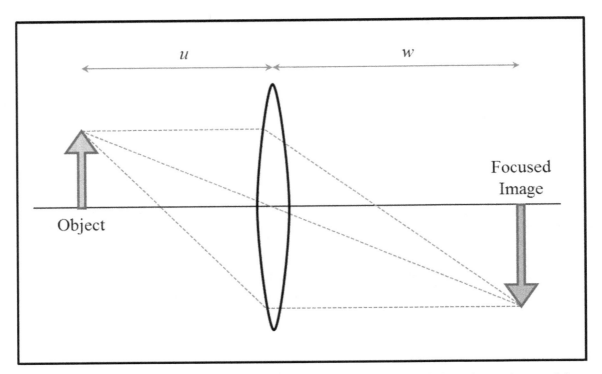

A schematic of the path of light beams traveling through a lens and forming a focused image. u represents the light source to lens distance; w is the lens to image distance

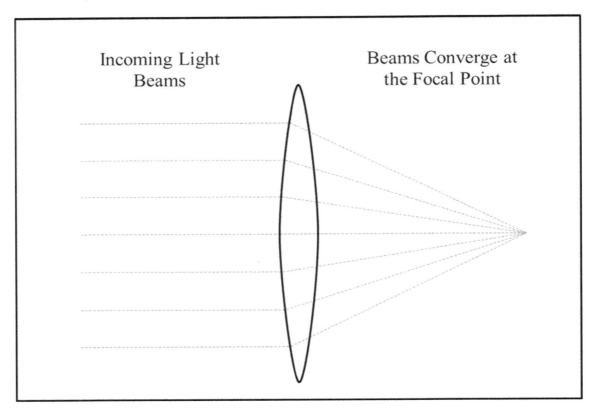

A schematic demonstrating the concept of a focal point; parallel beams of light from a very distant light source are refracted by the lens and converge at a single point some distance from the lens. That distance is the focal length, expressed by the variable f_{eye} for an eyepiece or f_{obj} for an objective lens.

PROCEDURE

Check that you have the following equipment: optical bench; two convex objective lenses; one convex eyepiece lens; two lens holders; three optical bench riders; glass projection screen.

Place the light source onto the optical bench, centering it as closely as possible to 0.0 centimeters. Place the objective lens somewhere on the optical bench between 30 and 70 cm from the light source. The exact initial positioning is up to you. Place the screen on the far end of the optical bench. You likely will not see any projected image initially. Move the glass screen inward (toward the lens) until a bright, focused image appears projected onto the screen. Adjust the position of the screen until the image is as sharp as you can get it. Note the positions of the three components – light source, lens, and screen – and determine the distances u and w in the table provided for Trial 1.

The variable u is the distance from the light source to the lens. The variable w is the distance from the lens to the screen when the image is sharp and in focus. Use those measurements to calculate $1/f$ and the focal length of the objective.

Next, move the objective to a new position at least 5 centimeters away from your initial position (either closer to or further from the light source). Move the screen and find the point where the image is again sharp and focused. Note the new positions and record u, w, $1/f$, and f for Trial 2. Repeat this procedure for two other unique positions of the objective. After the four trials are complete, average your focal lengths to determine the focal length of your first objective lens.

Repeat this measurement process for your second objective and the eyepiece lens.

Name _____ Id _____

Due Date _____ Lab Instructor _____ Section _____

Worksheet # 1

Focal Length of Objective Lens 1

Trial	u (in cm)	w (in cm)	$1/f$	f_{obj} (in cm)
1				
2				
3				
4				

Average focal length: [   ] cm

Focal Length of Objective Lens 2

Measurement	u (in cm)	w (in cm)	$1/f$	f_{obj} (in cm)
1				
2				
3				
4				

Average focal length: [   ] cm

Focal Length of the Eyepiece Lens

Measurement	u (in cm)	w (in cm)	$1/f$	f_{eye} (in cm)
1				
2				
3				
4				

Average focal length: [   ] cm

Name _____ Id _____

Due Date _____   Lab Instructor _____   Section _____

Postlab Questions

For each of the following questions, include all work, equations, and proper units.

1. Considering your objectives and eyepieces, what is the maximum magnification you could achieve by arranging them as the components in a simple refracting telescope?

2. Using those lenses from above, what is the minimum length for this telescope?

3. Building a simple refracting telescope: Point your optical bench toward a distant point at least 20 feet away. Place the objective lens in a lens holder at the 0.0 centimeter mark on the optical bench. Place the eyepiece on the optical bench and make sure it aligns with the objective. Look through the eyepiece-objective and move the two lenses closer or further apart until a sharp, magnified image appears. What is the separation between the two lenses?

4. How close is that separation to the calculated minimum telescope length from question 2?

5. Seeing Saturn's rings requires a magnification of at least 30×. Keeping your current eyepiece, what focal length would your objective need to see Saturn's rings? What is the minimum length of that telescope?

UNIT 3.2   TELESCOPES

OBJECTIVE
To explore the abilities and characteristics of major telescopes – past, present, and future

INTRODUCTION
Since Galileo Galilei's first optical astronomical telescope was used to study the phases of Venus and craters on the Moon, telescopes have evolved profoundly, becoming larger, more sensitive, and more powerful. Even today's largest telescopes still operate on the same basic principle: a large, light gathering mirror (or, less frequently today, a lens) gathers light from distant objects and focuses it into a sharper, brighter image which can be recorded and studied in detail. The curved, reflective surface of a mirror redirects incoming light from across its entire surface and channels it either to an eyepiece or a camera recording system. This main light-gathering optical system is called the telescope's *objective*. The larger the objective, the more fine details can be seen from an object, referred to as the telescope's resolution.

The objective is the telescope's largest surface area, capable of intercepting a large number of incoming beams of light and focusing them into a brighter, concentrated point. The spot where the beams of light came together is called the *focal point* (the single point at which all of the arriving light was focused) as shown in the diagram of Unit 3.1. The distance from the mirror or lens to the spot where the beams of light converge into a sharp, focused point is called the *focal length*, and is a fixed, physical characteristic of the lens or mirror, based on the size, shape, and composition of the objective.

On its own, the objective was not useful for astronomical viewing. The converged light beams would have to be reoriented and directed through the observer's pupil. This was accomplished by the eyepiece lens, a smaller lens that intercepted the light from the objective and focused it into the observer's eye. Eventually, with the birth of film and eventually electronic photography, the eyepiece was eliminated, replaced with a more sensitive camera system. In the photography age of the telescope, the "speed" of the telescope became an important consideration. The speed of a telescope – or the focal ratio, sometimes also referred to as the "f-number" – is a measurement which indicates the image brightness that could be obtained from a photographic exposure. The smaller the focal ratio or f-number, the "faster" the telescope and the brighter the image would appear after an exposure. A slower telescope (with a larger focal ratio f-number) could produce an equally bright image, but the exposure time would be considerably longer.

For a faster telescope, images can quickly be recorded on a CCD or a film camera. A larger focal ratio results in a fainter image that takes a longer time to record on camera. Focal ratio is expressed as f/#, where # represents the ratio of objective focal length to objective diameter. For example, f/2 is considered a fast telescope and f/8 is considered a slow one. The focal ratio, which has no units, can be calculated by dividing the telescope's focal length by the diameter of the primary objective (so those measurements must be expressed in the same units).

The magnitude scale is an astronomical method for comparing the brightness of various stars, as they appear to the eye. It is a scale which runs backwards, with a lower number indicating a bright star. The naked human eye may only see objects with a magnitude smaller than 6. In total, there are approximately 9000 stars in the entire sky with magnitudes of 6 or less, the brightest, most visible nighttime star being Sirius. Telescopes, with their ability to gather faint light and focus it into a brighter image, allow humans to see objects which would otherwise be below the human vision threshold. This is referred to as the "limiting magnitude" and is the dimmest star that a human could see through a particular telescope. For example, a small pair of binoculars can make stars of magnitude 7 and 8 visible to human vision, allowing observers to see an additional 10,000 night sky stars as well as the faint light of the planets Uranus and Neptune, both of which are out of naked eye human vision range.

The larger a telescope's main lens or mirror, the fainter the star that can be seen or photographed. This is directly related to the telescope's "light gathering power," which is the amount of light two optical instruments gather when compared to one another. The larger a telescope's objective surface area, the greater the light gathering power, and may be expressed as a ratio of the area of a large telescope compared to the surface area of a smaller telescope. Engineers build larger and larger telescopes to take advantage of the greater light gathering power that can only be offered by larger and larger telescopes. For this reason, future telescopes – like the LSST, the Thirty Meter Telescope, and the Extremely Large Telescope – incorporate large light gathering surface areas into their design.

EQUATIONS AND CONSTANTS

Equation	Expression	Variables
Resolving Power	$R = \dfrac{11.6}{D}$	R : resolution in arcseconds D : diameter of objective in centimeters
Surface Area	$A = \pi \left(\dfrac{D}{2}\right)^2$	A : surface area of the objective D : diameter of the objective
Focal Ratio	$f/\# = \dfrac{f}{D}$	$f/\#$: speed or focal ratio of the lens or mirror f : focal length of the lens or mirror D : diameter of the lens or mirror
Limiting Magnitude	$m = 2 + 5\log(10 \times D)$	m : the dimmest magnitude detectable D : diameter of the lens or mirror in cm
Light Gathering Power	$LGP = \left(\dfrac{D_1}{D_2}\right)^2$	LGP : ratio of light gathering power D_1 : diameter of the larger objective D_2 : diameter of the smaller objective
Light Gathering Power	$LGP = \dfrac{A_1}{A_2}$	LGP : ratio of light gathering power A_1 : surface area of the larger objective A_2 : surface area of the smaller objective

PROCEDURE

Fill in the worksheet table based on the given data and using the equations above. Make sure to be careful with units where necessary. For example, the resolution equation always requires the diameter of the objective in centimeters, so make sure to convert if you are given a diameter in another unit, such as meters.

Name _____ Id _____

Due Date _____ Lab Instructor _____ Section _____

Worksheet # 1

Telescope Name	Diameter (cm)	Focal Length (cm)	Focal Ratio (f/#)	Resolution (arcsec)	Surface Area (cm²)	Limiting Magnitude
Galileo's First Telescope	3.9	99				
Herschel's Great Telescope	120	1200				
UTA Telescope	40	400				

The Large Research Telescopes

Telescope Name	Diameter (cm)	Focal Length (cm)	Focal Ratio (f/#)	Resolution (arcsec)	Surface Area (cm²)	Limiting Magnitude
Hubble Space Telescope	240	5760				
Very Large Telescope	820	1440				
Keck Telescope	1000	1750				

The Future and Planned Telescopes

Telescope Name	Diameter (cm)	Focal Length (cm)	Focal Ratio (f/#)	Resolution (arcsec)	Surface Area (cm²)	Limiting Magnitude
James Webb Telescope	650	1320				
Thirty Meter Telescope	3000	45000				

Data obtained from:

http://hubblesite.org

https://www.lsst.org

https://www.eso.org/sci/facilities

http://www.tmt.org/

Name _____ Id _____

Due Date _____   Lab Instructor _____ Section _____

Worksheet # 2

1. While Galileo's first telescope was small and primitive by today's standards, it was a massive observational leap for its time. The human pupil (which works like the objective lens of the eye) has a maximum diameter of 0.8 cm. How much more light gathering power did Galileo's telescope has than the naked human eye?

2. The European Space Agency had planned on building a massive telescope, referred to as the OWL (Overwhelmingly Large Telescope). The OWL would have been equipped with a 100 meter wide objective mirror. The OWL was cancelled to make way for the more affordable ELT (Extremely Large Telescope) with its 39 meter primary mirror. How much more light would the OWL be able to gather compared to the ELT?

3. What would be the resolution (smallest angle visible) of the OWL?

4. Which telescope is "faster," Herschel's Grand Telescope, built in 1785 or the Hubble Space Telescope, built in 1990?

5. Which of the following very dim objects could Galileo have spotted with his first telescope (check each that could have been seen)?

   ☐ Uranus, with a magnitude of 6.05

   ☐ Ceres, the largest asteroid in the Asteroid Belt, with a magnitude of 7.1

   ☐ Neptune, with a magnitude of 9.5

   ☐ Pluto, with a magnitude of 14.3

6. To the naked eye, the star Capella appears to be one yellowish star. However, through a telescope, it appears as two separate stars orbiting one another. Explain what power of a telescope – speed, light gathering power, or resolution – allows an astronomer to see Capella as separate stars.

7. Name at least two advantages you can think of for launching a telescope into Earth orbit. How might a smaller Earth-orbiting telescope be more advantageous than a very large ground based telescope?

UNIT 3.3   ASTROPHOTOGRAPHY

OBJECTIVE

To understand how images of the night sky objects are recorded

INTRODUCTION

Astrophotography is the art of photographing the night sky, either large, sweeping star fields or very close cropped deep sky objects, like galaxies and nebulae. With the proliferation of cheap digital imagers and reliable computer controlled telescopes, amateur astrophotography has become popular and widespread.

An important aspect of photography is exposure time. Exposure time describes the duration of light collection by your camera. Human vision is not at all like a camera, despite the human eye sometimes being compared to a biological camera. The processes light by pumping chemicals which conduct electrical signals to the brain, creating a continuous moving image with varying sharpness and contrast and sensitivity. Photographs capture the light created by an event over some set time period, and photographers must balance between the need for light and the need for sharpness, and that involves determining an exposure time.

The longer the exposure, the more light that is recorded and the brighter the subject appears. Too long of an exposure and the incoming light will "spill" across your recording device, overexposing the image and whiting out all details. In addition, any motion from the subject will be recorded as well, blurring the fine details. For this reason, old Victorian era photographs from the 1800s usually featured very staid, unsmiling subjects. This was prompted by the subjects of the photo having to hold very still for a five minute exposure, which makes holding a smile difficult without moving and blurring the picture. On the other hand, too short of an exposure and the subject will be underexposed and dark.

A CCD – or charge coupled device – is a specialized light sensor developed for recording astronomical images. Digital imaging revolutionized telescopic studies, as their sensitivity allow for the collection of even the faintest light and extremely long exposures

The focal ratio – also called the f-number or f-stop – is an important measure of a camera's ability to gather light. The f-stop is a ratio between the diameter of a lens and the distance from the lens where the light beams are focused. The smaller the ratio, the more light enters the camera and reaches the recording surface over a given time period. For a "fast" camera – with a focal ratio of f/2, for example – a bright image can be obtained with a relatively short exposure of a few seconds or several minutes. A "slow" camera – with a focal ratio of f/8 or f/10 or f/16, for example – can produce an equally bright image but will require a considerably longer exposure time for the photograph. An f/2 camera may produce a bright image of a nebula in 15 minutes. An f/16 camera could produce an equivalent photograph but would require 8 times the exposure (a 120 minute or 2 hour exposure).

When you look at an object, you judge its size by how much of your vision the object takes up. An object may fill a large portion of your vision due to a large physical size, due to being close to your eye, or some combination of those two. This apparent size is referred to as the angular size, and is dependent on both the object's size and distance. Measuring an object's angular size is generally easy to accomplish, either using surveying equipment or knowing the size of the field of view of the camera. Actual distances and actual sizes in space are generally more difficult to measure but so long as one of those values can be determined – size or distance – the other value is easy to calculate.

EQUATIONS AND CONSTANTS

Equation	Expression	Variables
Angular Size	$$\theta = \frac{L \times 206265}{S}$$	θ : apparent size of an object in arcseconds S : the actual size of the object being observed L : the actual distance between object and observer
Focal Ratio	$$f/\# = \frac{f_0}{D}$$	$f/\#$: speed or focal ratio of the lens or mirror f_0 : focal length of the lens or mirror D : diameter of the lens or mirror

Name _____ Id _____

Due Date _____ Lab Instructor _____ Section _____

Worksheet # 1

1. Use an optical system (camera, telescope, binoculars, etc.) to photograph 3 separate objects visible in the night sky. Using a source (like Stellarium) which gives up-to-date information, record pertinent information about the object, such as its true size and magnitude (making sure to state what units you used). Include – either in print out or electronically – the images, making sure to label the name of the object. Using the distance and actual size, calculate an angular size for each object.

Object	Distance	Actual Size	Magnitude	Angular Size

2. Record the diameter of primary objective, focal length, and focal ratio of instrument you used for observing each object. Also, note down the exposure time for each object observed.

Object	Focal Length	Diameter	Focal Ratio	Exposure Time

UNIT 3.4   SPECTRAL ANALYSIS

OBJECTIVE

To understand the basic principle of spectroscopy by comparing the spectra of unknown gases to those of known sources

INTRODUCTION

The light that we receive from the Sun, stars and other objects, when passed through a prism or diffraction grating, disperses into its constituent colors (or wavelengths) resulting in a spectrum. The scientific study of objects based on spectral analysis is called spectroscopy.

Spectroscopy is one of the most valuable techniques of an astronomer. Although spectroscopy is a complex art, most of our knowledge about stars with respect to their chemical composition, temperature, velocity, etc. has been gained through spectroscopy. There are fundamentally three types of spectra. They were first classified by the German physicist Gustav Kirchhoff in 1859.

Continuous spectrum: A hot glowing solid or a high density hot gas (e.g. the common light bulb) produces all the wavelengths in the spectrum. Such a spectrum is called a continuous spectrum because there are no gaps or breaks between the colors. A rainbow is an excellent example of a continuous spectrum.

Emission spectrum: A low density gas when heated produces spectral lines corresponding to only certain wavelengths that represent the characteristics of the gas. This type of spectrum is called emission spectrum and shows bright lines on black background.

Absorption spectrum: When the light from a continuous spectral source passes through a cooler gas, this gas absorbs certain wavelengths resulting in spectrum with dark lines on the continuous background. This type of spectrum is called an absorption spectrum. The dark lines represent the characteristics of the cooler gas.

The emission and absorption lines are related to the electron orbitals for each element and the presence (or absence) of different spectral lines are very indicative of the atoms which are present in the gas. Every element existent in the Universe has its own electron orbits that are different from those of any other element. Therefore, each element produces a unique spectrum, also called a fingerprint or spectral signature of the element.

PROCEDURE

In this lab exercise you will observe continuous, emission and absorption spectra through a spectroscope. A spectroscope is an instrument that has a narrow slit on the side facing the light source on one end and a diffraction grating (a glass or plastic plate with thousands of lines grooved on it) that acts as a dispersing device on the observer's end.

Part 1: You will identify unknown gases by noting down their wavelengths on the chart and matching them with the wavelengths of known sources.

A. You will have at your work station several labeled bulbs called gas discharge tubes. In these tubes is a very thin gas which is heated by the black box power supply until the electrons are excited and begin to produce light. The bulbs may have strikingly different colors when powered on, indicative of the element that makes up the heated gas sample.

B. Be careful with the equipment. The power boxes carry a large electric charge, the bulbs may heat to a few hundred degrees, and the glass stems are fragile and prone to snapping if mishandled. Listen to and follow all lab instructions. The equipment is expensive and fragile! So, it is pivotal and mandatory that you understand the rules for their use.

C. Together with your lab partner, power up each of the discharge tubes. Make sure the eyepiece of the spectrometer is aligned with and close to the stem of the bulb, allowing maximum light from the source to enter the spectrometer. If you do not see bright, prominent spectral lines or light, pivot the eyepiece slowly to fix the alignment with the bulb. The spectral lines will be bright and unmistakable.

D. Carefully, using the scale on the spectrometer and the worksheet graphs, mark the position of each spectral line. Using a colored pencil to match the spectral line color may prove helpful in determining the gas's elemental composition. Where there is a space for "Bulb Label" on the worksheet, identify the bulb by its tag.

E. Some lines will be brighter and more prominent than others. Under the "Prominent Visible Lines" header, write down the general color and measured wavelength of the brightest spectral lines, or describe the spectra in general (i.e., bright blue hue to bulb; many closely spaced red lines of equal brightness; very broad, smeared lines; etc.)

F. Using your sketch of the spectra, identify the likely element that you are looking at.

Part 2: The scale model of the hydrogen atom. The prominent colored spectral lines of hydrogen are caused by electrons dropping from higher energy levels and settling into the second energy level. For example, the bright red photons with wavelengths of 656 nm are caused by electrons falling from energy level number 3 into energy level number 2. A drop from level 4 to level 2 creates a higher energy photon with a teal color to the human eye. A drop from 5 to 2 creates an even more energetic photon.

Using the equation for energy level distance and the wavelength (λ) of the emitted photons, determine the actual size of the energy levels and then draw a scale model of the hydrogen atom. The proton is shown on worksheet 2 as a small black dot. The gray circle represents the hydrogen atom's second energy level (not pictured is energy level 1, which, on the given scale, would be extremely close to the nucleus).

When drawing the energy levels, use the scale that 1 real-world nanometer = 5 centimeters on the scale. Mark off the distance to each energy level (measuring the distance from the nucleus) and use a compass to draw a circular electron orbit around the nucleus. This represents the spacing of four orbits around hydrogen.

EQUATIONS AND CONSTANTS

Equation	Expression	Variables
Distance of an Energy Level from the Nucleus	$r_n = \dfrac{\lambda}{(4.73)\lambda - 1630}$	r_n: the radius of the electron orbit in nm (i.e., r_3 for the 3rd energy level, r_4 for the 4th energy level) λ: the wavelength of an emitted photon in nm

Name _____ Id _____

Due Date _____ Lab Instructor _____ Section _____

Worksheet # 1

Bulb 1   Bulb label _____

400   500   600   700

Nanometers

Prominent visible lines: **Element**

Bulb 2   Bulb label _____

400   500   600   700

Nanometers

Prominent visible lines: **Element**

Continue....

Bulb 3   Bulb label _____

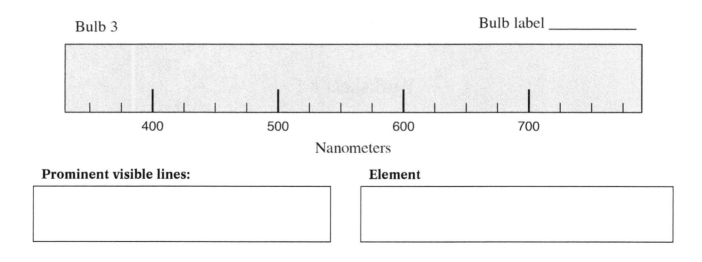

Nanometers

Prominent visible lines: **Element**

Bulb 4   Bulb label _____

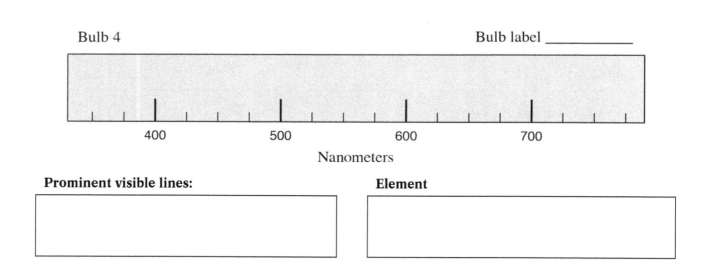

Nanometers

Prominent visible lines: **Element**

Name _____ Id _____

Due Date _____   Lab Instructor _____   Section _____

Worksheet # 2

Color of Emission Line	Energy Level	Wavelength of Photon in nm	Radius of Orbit in nm	Scaled Radius of Orbit in cm
Red	3	656		
Teal	4	486		
Violet	5	434		

Scale model of the second energy level and the hydrogen nucleus (already drawn) enlarged by 50 million times

Scale: 1 nm 5 cm

UNIT 4: A CLOSER LOOK THROUGH TELESCOPES

UNIT 4.1   LUNAR FEATURES

OBJECTIVE
To observe the Moon, its phases and identify large, key features on the lunar surface

INTRODUCTION
As the closest celestial object to the Earth, our Moon, is one of the most easily observed object in the sky (and remains the only extraterrestrial body that humans have stepped foot upon so far). Due to a process called "tidal locking," the Moon only reveals one hemisphere – the "near side" – to the Earth. The near side of the Moon is covered by familiar features: the large, dark lunar *maria* (Latin for "seas"; singular: *mare*) and the light colored, silvery lunar highlands. With the naked eye, the Moon shows little in the way of geological features, such as craters, valleys, and mountain chains.

From the northern hemisphere of the Earth, the Moon's right side (its eastern hemisphere) is dominated by continuous, light colored highlands in the south and several round, mostly separate dark mare in the north (namely, the Sea of Serenity, Sea of Tranquility, Sea of Crisis, and Sea of Fertility). The Moon's western hemisphere (its left side, as seen from the northern hemisphere of the Earth) is dominated by large, dark, overlapping maria that form a hemisphere-wide plain. They constitute ancient lava plains, formed when active lunar volcanoes sent cascades of liquid rock (magma) into low lying basins.

At 3475 kilometers in diameter, the Moon is just over one-quarter the diameter of the Earth, equating 1.2% of Earth's mass. This is the smallest size discrepancy between any planet and its moon in the Solar System. Our Moon, airless and exposed to the vacuum of space, shows a long history of geological activity and collisions with asteroids, meteoroids, and comets. Vast systems of craters leave the Moon visibly battered and scarred. Ray craters are those with long streaks of violently ejected material emanating from the central crater. In these instances, the impact with the lunar surface scooped up huge amounts of rock and dust and blasted it over a wide, sometimes thousands of square mile area.

PROCEDURE
Naked eye observations: for this part, notice the phases of the Moon, including the amount of the near side of the Moon which is in darkness and which hemisphere is actually visible. Using the outline of the Moon provided on Worksheet #1, draw a line to indicate the location of the terminator (the day-night dividing line running from the Moon's north pole to its south pole). If the Moon is full, note that it is a full Moon and fully lit. Utilizing a moon map with named lunar feature, label the location of the major lunar regions of Mare Serenitatis, Mare Tranquillitatis, Mare Imbrium, and Mare Procellarum.

Using a telescope: for this section, you will need to plan ahead of time on the targets you wish to observe. You will use a telescope and observe the Moon in much greater detail. Record the date, time, the telescope in use (its diameter, such as 8 inches, 10 inches, etc.) and the magnification of the telescope. Scan the lunar surface for the following features: 3 craters (at least one with a ray system); 2 maria; and 1 mountain range. There are 6 spaces provided for sketches, though you may identify more than one feature per field of view (in other words, one sketch may show a ray crater as well as a standard crater or a maria and a mountain range). Use the table of prominent and easily identifiable lunar features (Datasheet #1) to target specific areas of the Moon and specific features to be observed.

In the circle provided on Worksheets #2 and #3, sketch exactly what you see in the telescope's eyepiece, drawing craters, valleys, ray systems, and shadows as you see them. Label the name of a particular crater, mountain range, maria, or other interesting feature that you observed.

DATASHEET #1

LUNAR FEATURES		
Craters	**Mare**	**Mountains**
Langrenus	Mare Crisium	Caucasus
Aristoteles	Mare Foecunditatis	Apennines
Eudoxus	Mare Nectaris	Alps
Hercules	Mare Tranquillitatis	
Atlas	Mare Serenitatis	
Theophilus	Mare Imbrium	
Piccolomini	Mare Nubium	
Hipparchus	Mare Humorum	
Plato	Oceanus Procellarum	
Archimedes		
Copernicus		
Kepler		
Aristarchus		
Eratosthenes		
Tycho		

Name _____   Id _____

Due Date _____   Lab Instructor _____   Section _____

Worksheet # 1

Name _____ Id _____

Due Date _____ Lab Instructor _____ Section _____

Worksheet # 2

E W	Observation Date:	
	Time:	
	Instrument:	
	Magnification:	
	Identified Features	
	Craters:	
	Maria:	
	Mountains:	
	Other:	

E W	Observation Date:	
	Time:	
	Instrument:	
	Magnification:	
	Identified Features	
	Craters:	
	Maria:	
	Mountains:	
	Other:	

Name _____ Id _____

Due Date _____ Lab Instructor _____ Section _____

Worksheet # 3

Observation Date:	
Time:	
Instrument:	
Magnification:	
Identified Features	
Craters:	
Maria:	
Mountains:	
Other:	

E   W

Observation Date:	
Time:	
Instrument:	
Magnification:	
Identified Features	
Craters:	
Maria:	
Mountains:	
Other:	

E   W

UNIT 4.2   MERCURY, VENUS, MARS AND SATURN

OBJECTIVE
To locate, observe, and sketch some of the Solar System's naked-eye visible planets

INTRODUCTION
These four planets – named after the Roman gods of speed, beauty, war, and time, respectively – are visible planets, which can easily be spotted at night when they are above the horizon.

MERCURY: This small, fast-moving planet is only visible either shortly before sunrise or shortly after sunset, and rarely for more than an hour-and-a-half before the Sun rises or sets. Not much larger than the Earth's Moon and often lost in the glare of the Sun, Mercury only displays a small disk that undergoes Moon-like phases when observed with a telescope. No other features of this cratered planet are visible.

VENUS: No other planet appears as bright as Venus, hence its ancient association with the goddess of beauty. It can be seen for a maximum time of about 3 hours as an "evening star" after sunset or as a "morning star" before sunrise. At its brightest, Venus can even be seen during the day time.

Much like the Moon and Mercury, Venus displays phases. When Venus is closest to Earth (on the same side of the Sun as our planet) it will display a large crescent phase. On the far side of its orbit it will appear as a small, intensely bright gibbous disk. Because Venus is wrapped in a thick, featureless, unbroken cloud deck of highly reflective carbon dioxide, no surface features of Venus are visible.

MARS: Mars's orange-red color against the black night sky makes it easy to pick out amongst the stars. Mars is a desert planet covered in iron rich dust, hence the red-amber color. At its closest approach to the Earth and under good viewing conditions, you may be able to see a polar icecap and some dark markings of Mars's vast equatorial deserts. Due to its small size, Mars never appears as large as Jupiter and Saturn despite being far closer to Earth than those two.

SATURN: The furthest planet visible to the naked eye, this slow moving gas giant planet circles the Earth once every 29 years, earning its naming as the elderly, hobbled father of the gods of the Roman pantheon. Saturn's hydrogen-helium clouds appear gold and cream colored in a small telescope, with some subtle hint of striped cloud decks. Saturn's most spectacular feature is its rings, a system of highly reflective orbiting ice particles stretching almost 300,000 kilometers from end to end. With a moderately large telescope, the Cassini division gap in the rings can easily be seen.

Since Saturn is tilted, much like Earth, its rings appear inclined to our line of sight. However, during some years, they may appear edge-on and nearly invisible. Other years, Saturn may be tilted toward our planet, tilting its impressive ring system toward Earth. Some of Saturn's moons can also be seen, particularly the large, icy, atmosphere-enshrouded moon Titan. Given Saturn's axial tilt, the moons do not appear to align with Saturn and one another, as Jupiter's Galilean moons do. For most of the time, however, they are not "in line" as are Jupiter's moons.

PROCEDURE

1. Locate the planet assigned for observation in the sky, using a star chart, online source, or planetarium software. Determine what constellation the planet will appear in front of that evening.

2. Using star charts, sketch the constellation in the space provided on the worksheet (under the heading "Constellation") and mark with an (x) the position of the planet in that constellation.

3. Observe the planet through a telescope, making note of the diameter of the telescope's mirror or lens, its magnification, and the date of observation in the space provided.

4. Under the heading "Planet" make a sketch of the planet as you see it within the circle (the circle representing the field of view of the sky through telescope's eyepiece).

Example: If Galileo Galilei were to have used one of these worksheets on the night of his discovery of the moons of Jupiter, it would have appeared as follows:

		Constellation	Planet
Planet name:	*Jupiter*		
Constellation name:	*Taurus*		
Date:	*Jan 7, 1610*		
Telescope diameter:	*3.6 cm*		
Magnification:	*9x*		

Name _____ Id _____

Due Date _____ Lab Instructor _____ Section _____

Worksheet # 1

		Constellation	Planet
Planet name:			
Constellation name:			
Date:			
Telescope diameter:			
Magnification:			

		Constellation	Planet
Planet name:			
Constellation name:			
Date:			
Telescope diameter:			
Magnification:			

		Constellation	Planet
Planet name:			
Constellation name:			
Date:			
Telescope diameter:			
Magnification:			

Name _____ Id _____

Due Date _____ Lab Instructor _____ Section _____

Worksheet # 2

		Constellation	Planet
Planet name:			
Constellation name:			
Date:			
Telescope diameter:			
Magnification:			

		Constellation	Planet
Planet name:			
Constellation name:			
Date:			
Telescope diameter:			
Magnification:			

		Constellation	Planet
Planet name:			
Constellation name:			
Date:			
Telescope diameter:			
Magnification:			

UNIT 4.3   JUPITER AND THE GALILEAN MOONS

OBJECTIVE

To observe the planet Jupiter, its four major moons, and note their changing positions

INTRODUCTION

Jupiter and its major moons played an important role in overturning the long-held ancient belief in an Earth-centered or geocentric Universe. To the naked eye, Jupiter appears as a single bright point of light moving against the stars positioned along the ecliptic, following much the same repeating path as the Sun, Moon, and other major planets of the Solar System. Given Jupiter's great distance and slow revolution, it requires just about 11 years to complete one orbit.

Even through a small telescope, Jupiter reveals a wide array of features, from its bigger than Earth-sized Great Red Spot to its alternating bands of maroon and cream colored cloud bands. Most stunning to Galileo and early Jupiter observers were Jupiter's four main moons, now called Galilean moons. Appearing as little more than pinpoints of light strung across the space surrounding Jupiter, Galileo at first mistook these objects for background stars which would vanish from view as Jupiter whisked across the ecliptic. However, repeated observations showed that these pinpoints were in fact following Jupiter, passing in front of and behind the planet. They were, in fact, satellites of the giant planet, dealing a blow to the geocentric model that everything in the Universe orbited the Earth.

Jupiter has at least 63 moons, ranging in size from small kilometer-wide captured asteroids to gigantic rocky/icy moons larger than the former planet Pluto. The four Galilean moons are named after mythological figures important to the Roman god Jupiter: Io, the innermost of the large Galilean moons; Europa, the moon which may have a warm ocean hidden beneath its icy crust; Ganymede, an icy/rocky moon with a diameter larger than the planet Mercury; and Callisto, the outermost moon. These moons have nearly circular orbits aligned with Jupiter's equator. Given Jupiter's very small axial tilt, these moons appear to move back and forth directly across Jupiter's equator.

PROCEDURE

This lab will require a number of observations of Jupiter, taken over the course of several days or up to two weeks. Using telescopes to make careful estimations of the positions of the moons of Jupiter over the course of the observations, you will be able to identify each of the Galilean moons.

Observe Jupiter through an astronomical telescope. As a test of the quality of your observations, be sure that at least one of Jupiter's dark cloud bands are visible. Those cloud bands will also help you determine the orientation of Jupiter's equator, as the bands align with Jupiter's equator and also align with the orbits of the Galilean moons.

Determine the location of the Galilean moons. You may see up to four of them strung out along either side of Jupiter. If you cannot see all four moons, one of them may be out of the telescope's field of view or – more likely – may be behind or in front of Jupiter and lost from view. Make a rough sketch of the field of view, determining which side of Jupiter the moons are on, east or west (also keeping in mind that telescopes tend to invert images, so that north is down and south is up, east is located on the right and west on the left). It is critical that you maintain the same orientation through the observations. Transfer your sketches to the spaces provided on Worksheet #1, careful to accurately estimate the distance between Jupiter and the various visible moons.

It is important that you use the same magnification from observation to observation, so make sure to record the magnification of the telescope you are using (making sure that the magnification is low enough to see the outermost moons whose orbits take them far from Jupiter). Make 7 observations over the course of no more than two weeks and track the motion of the moons.

You may make more than one observation on any given night. For example, you may observe one of the moons very close to Jupiter's disk during one observation and make a sketch. Some time later, you may notice that this moon has either moved noticeably farther away from the disk or moved closer to Jupiter. Moving closer, you may find that the moon has vanished, either in front of Jupiter and washed out by the planet's glare or behind Jupiter and lost in its shadow. No matter the case, a few hours later you may see it re-appear again on the other side of Jupiter. When in front of Jupiter, you may see its shadow cast on Jupiter's cloudy surface, taking the form of a small black circle falling on top of the clouds. You can follow and track the moon's appearance and disappearance using planetarium or night sky software, such as Stellarium.

After making 7 observations, the motions of the various moons should be clear. Some physical parameters of the moons' orbits – greatest distance from Jupiter's center in terms of Jupiter diameters and the orbital period in days – are listed below. Using this information and your sketches, on Worksheet #1, identify each moon by labeling it in your sketch with its Roman numeral symbol.

Symbol	Name	Distance from Jupiter (in km)	Distance from Jupiter (in Jupiter diameters)	Sidereal period (in days)
I	Io	420,000	3.0	1.77
II	Europa	670,000	4.8	3.55
III	Ganymede	1,100,000	7.9	7.14
IV	Callisto	1,900,000	13.6	16.69

Example of data set:

Date: *Jan 7*	
Time: *8:44 PM*	

Name _____ Id _____

Due Date _____   Lab Instructor _____   Section _____

Worksheet # 1

| Telescope diameter: | |
| Magnification: | |

Date:	
Time:	
Date:	
Time:	
Date:	
Time:	
Date:	
Time:	
Date:	
Time:	
Date:	
Time:	
Date:	
Time:	

UNIT 4.4 DOUBLE STARS, CLUSTERS AND GALAXIES

OBJECTIVE

To locate and observe some of the night sky's deep space objects, encompassing double stars, open and globular clusters, nebulae, and galaxies

INTRODUCTION

To the naked eye on a clear, dark night, the stars in the night sky look like single, isolated points of light. This was the sky of the ancient sky watchers of, e.g., Babylon, Egypt and Greece, where little more than stars of varying brightness could be seen. However, the advent of telescopes showed the sky to be dotted with interesting, beautiful, and amazing sights. Even with binoculars or a small telescope some of these objects become clearly visible. Double stars are closely spaced stars which can be separated into two separate points of light. The nearness of double stars is either coincidental, where two very distant stars simply appear along the same line of sight, or due to the stars actually being gravitationally bound to one another. On the other hand, open clusters are swarms of young, related stars which have as of yet not dispersed into the Milky Way Galaxy and are still loosely bound to nearby stars. Globular clusters are immensely old star clusters, sometimes containing hundreds of thousands of ancient stars born together billions of years ago and orbiting about the Milky Way Galaxy.

Nebulae – Latin for "clouds" – are immense hydrogen-helium rich structures. Nebulae may be either the raw building blocks for newborn stars – such as the Great Nebula in Orion – or the remains of highly evolved stars which have ended their stellar life times, like the Crab Nebula or Cat's Eye Nebula. Finally, galaxies are gargantuan, gravitationally bound masses of stars, dust, gas, planets and stellar remnants stretching tens of thousands of light years across. The Milky Way is just one of countless galaxies populating the observable Universe.

The tables below contain some of the most easily observable night sky objects, which may be clearly resolved with a telescope only a few inches in diameter. Objects in circumpolar constellations – i.e., always above the horizon – are listed first, followed by seasonal objects visible in the spring, summer, autumn, and winter sky.

PROCEDURE

The data tables included in this lab have a variety of bright objects and stars which allow for reliable telescopic or binocular observations. The objects are numbered 1 through 45. Determine which objects you will be viewing and record the pertinent information: Object Number, Type of Object (double star, nebula, galaxy, etc.), and location within its constellation. Refer to the star charts in this book, an online source, or planetarium software to determine where exactly the object falls within its constellation.

Using the manual's star charts (located in the appendix) as reference, sketch the constellation and mark the position of the object. Use a small (x) to designate the object's position.

Observe the object through a telescope and make a sketch of the object. As reasonably as possible, plot the position of visible stars in the telescope's field of view. For nebulae or galaxies, shade in the visible structure and shape.

Fill in the required information for the telescopic observation, including the date of observation. Some objects are best seen through binoculars, as, e.g., the Pleiades. Write "binocular" on the line reserved for the telescopic information.

When you observe double stars, you will note that, typically, one of the components is brighter than the other. Use larger dots for brighter stars. Also, the colors of the components may be noticeably different from each other. If observed, note the colors.

Example:

		Constellation	Sketch Object
Object #:	46 – The Hyades Cluster		
Constellation name:	Taurus		
Date:	Jan 1		
Object Type:	Open Cluster		
Telescope diameter:	5 in		
Telescope magnification:	20x		

DATASHEET #1

DOUBLE STARS

Object #	Constellation	Other Name	Apparent visual magnitude of components	Separation in arcseconds
1	α in Ursa Minor	Polaris	2; 9	18
2	ζ in Ursa Major	Mizar	2.4; 4	14
3	δ in Cepheus		3.7 – 4.4; 7.5	41
4	α in Canes Venatici	Cor Caroli	3.2; 5.7	20
5	γ in Leo		2.3; 3.5	5
6	β in Cygnus	Albireo	3; 5	35
7	ε in Lyra		4.2; 5.5	150
8	α in Hercules		3.5; 5.5	5
9	β in Scorpius		3; 5; 5	1–14
10	γ in Andromeda		2; 5	10
11	α in Capricornus		4; 4	375
12	α in Gemini	Castor	2; 3	5
13	β in Monoceros		5; 5; 6	7–10

Adapted from "Double Stars, Open and Globular Clusters, Nebulae, and Galaxies" from Practical Astronomy: Observations, Experiments, and Exercises Revised 2nd Edition by U.O. Herrmann & B.C. Thompson. Copyright © 1996 by Kendall Hunt Publishing Company. Reprinted by permission.

DATASHEET #2

STAR CLUSTERS, NEBULAE, AND GALAXIES

Object #	Name	Constellation	Notes (m$_v$ is the visual magnitude of the object)
14	M 81, M 82	Ursa Major	These two spiral galaxies are only 0.5° apart and may be seen in the same field of view. M81 is very elongated. m$_v$: 8 and 9
15	M 52	Cassiopeia	Open cluster, m$_v$: 7.3
16	NGC 663	Cassiopeia	Open cluster
17	NGC 7789	Cassiopeia	Open cluster
18	M 44	Cancer	Praesepe or Bee-Hive open cluster, visible with naked eye, best viewed through binoculars
19	M 3	Canes Venatici	Globular cluster, m$_v$: 6.3
20	M 51	Canes Venatici	The well-known, often photographed Whirlpool galaxy, m$_v$: 8
21	NGC 5139	Centaurus	Large globular cluster, not visible at latitudes much higher than 35° N
22	M 13	Hercules	The famous and very bright Hercules cluster, m$_v$: 5.7
23	M 57	Lyra	The famous ring nebula, planetary nebula, usually described as a smoke-ring or a doughnut. The center star is visible only through large telescopes, m$_v$: 9.3
24	M 27	Vulpecula	The Dumbbell nebula. Another planetary nebula. On clear, dark nights this appears extremely impressive, m$_v$: 7.6
25	M 8	Sagittarius	Lagoon nebula, diffuse nebula, visible to naked eye
26	M 17	Sagittarius	Horseshoe nebula, diffuse nebula
27	M 20	Sagittarius	Trifid nebula, diffuse nebula
28	M 22	Sagittarius	One of the finest globular clusters, m$_v$: 5.9
29	M 24	Sagittarius	Open cluster, m$_v$: 4.6
30	M 4	Scorpius	Very compact globular cluster, m$_v$: 6.5
31	M 6, M 7	Scorpius	A pair of open clusters, both visible simultaneously with wide-angle binoculars, m$_v$: 5.5
32	M 80	Scorpius	Globular cluster, m$_v$: 8
33	M 31	Andromeda	The great Andromeda galaxy, the only spiral galaxy visible to the naked eye, 2 mill. light-years distant, m$_v$: 4.8
34	M 32	Andromeda	An elliptical galaxy, companion to M 31, m$_v$: 9
35	M 34	Perseus	Open cluster, m$_v$: 5.5

(Continued)

Adapted from "Double Stars, Open and Globular Clusters, Nebulae, and Galaxies" from *Practical Astronomy: Observations, Experiments, and Exercises* Revised 2nd Edition by U.O. Herrmann & B.C. Thompson. Copyright © 1996 by Kendall Hunt Publishing Company. Reprinted by permission.

STAR CLUSTERS, NEBULAE, AND GALAXIES (Continued)

36	NGC 869, NGC 884	Perseus	The famous double cluster, use very low power to see both in the same field of view.
37	M 33	Triangulum	Spiral galaxy, m_v: 6
38	M 2	Aquarius	Globular cluster, m_v: 6
39	M 15	Pegasus	Globular cluster, m_v: 5.2
40	M 37	Auriga	Open cluster, m_v: 6
41	M 35	in Gemini	Open cluster, m_v: 5.3
42	M 1	in Taurus	Crab nebula, remnant of the 1054 supernova explosion, faint cloud, m_v: 8.4
43	M 45	in Taurus	Pleiades or Seven Sisters, open cluster. Best view through binoculars, m_v: 1.6
44	M 41	in Canis Major	Open cluster, m_v: 4.5
45	M 42	in Orion	Diffuse nebula, the Great Orion nebula, in the sword. Brightest of the diffuse nebulae. Look for "Trapezium", a quadruple star system in the nebula.

Name _____ Id _____

Due Date _____ Lab Instructor _____ Section _____

Worksheet # 1

		Constellation	Sketch Object
Object #:			
Constellation name:			
Date:			
Object type:			
Telescope diameter:			
Magnification:			

		Constellation	Sketch Object
Object #:			
Constellation name:			
Date:			
Object type:			
Telescope diameter:			
Magnification:			

		Constellation	Sketch Object
Object #:			
Constellation name:			
Date:			
Object type:			
Telescope diameter:			
Magnification:			

Name _____ Id _____

Due Date _____ Lab Instructor _____ Section _____

Worksheet # 2

		Constellation	Sketch Object
Object #:			
Constellation name:			
Date:			
Object type:			
Telescope diameter:			
Magnification:			

		Constellation	Sketch Object
Object #:			
Constellation name:			
Date:			
Object type:			
Telescope diameter:			
Magnification:			

		Constellation	Sketch Object
Object #:			
Constellation name:			
Date:			
Object type:			
Telescope diameter:			
Magnification:			

UNIT 5: A CLOSER LOOK WITH OTHER TOOLS

UNIT 5.1   THE ECLIPTIC

OBJECTIVE

To observe the changing position of the Sun against the background stars and to draw conclusions about the position of the Earth during the year

INTRODUCTION

The path that the Sun (and the planets and the Moon, though the Sun was most important because of its brightness) takes through the stars is called the ecliptic, and was repeated year after year after year. The constellations through which the Sun passes are known, collectively, as the zodiac. Even in modern culture we remember the Greek names of the zodiac constellations, which appear as the various "signs" (such as Sagittarius, the Archer, and Cancer, the Sea-Goat). The Greeks – whose reckoning of the sky and heavenly motions would profoundly influence astrologers from ancient Rome, ancient India, the later Islamic world, and the modern world – chose twelve constellations to outline the zodiac. As newspaper horoscopes readily show, it was believed that the stars of these constellations – along with the planets – could have profound effects on human beings' behaviors, fortunes, and destinies. This superstition is referred to as astrology, in contrast to the science known as astronomy.

The primary reason for the important nature of the zodiacal constellations lies not in anything particularly special about the stars themselves, but in their positioning in the sky. At some point during the year, the Sun will appear somewhere within the boundary of these historic twelve constellations. Ancient sky watchers could not accept that these constellations – through which the Sun traveled the sky – were not somehow extremely important to human affairs.

While we continue to call this the "path of the Sun" through the stars, the explanation for the march of the Sun is decided entirely by the position of Earth. As our planet revolves around the Sun, with the north pole pointing to Polaris, observers on Earth will see the Sun move from constellation to constellation (or, more precisely, we will see different constellations at night as the Sun stops washing out stars with its overwhelming light). So, when the Sun is in front of a constellation, such as Virgo, that means that a line drawn from Earth to Virgo will pass right through the Sun. During those weeks, Virgo will be right behind the Sun and impossible to view from the ground.

It should be noted that the IAU (International Astronomical Union) formalized the division of the celestial sphere into 88 official constellations, which was done in 1922 with the help of Henry N. Russell. Based on this concept, the number of constellations crossed by the Sun is identified as 13, rather than 12, with Ophiuchus being the "additional" constellation. This official determination of constellations should also be used for the following exercises.

PROCEDURE

Plotting the path of the Sun may be done in one of two ways: by naked eye observation of the Sun against the background stars or by using celestial coordinates. In either case, use the star charts provided in the following worksheets to plot the position of the Sun. If plotting by naked eye alone, using a planetarium, estimate the position of the Sun against the background stars, bright pointer stars to properly position the Sun. Plot each of the points on the provided star charts and connect them with a smooth, best fit curve showing the continuous path of the Sun through the celestial sphere. Use the star chart to estimate to the declination to the nearest degree and estimate the RA to the nearest hour and 10 minutes. Estimate the position of the Sun on your birthday and indicate that position on the ecliptic path with a star (*) and the calendar date.

If using a table of coordinates or software to track the Sun's coordinates, record the RA (in hours and minutes) and the Declination (in degrees) of the Sun in the chart first. Then plot those points on the star chart and connect them with a smooth, continuous best-fit line to create the ecliptic curve. Estimate the position of the Sun on your birthday and indicate that position on the ecliptic path with a star (*) and the calendar date.

On Worksheet #2, the small circle at the center represents the Sun, with the dashed line representing the orbit of the Earth and the background stars of the celestial sphere represented by the bold, outermost circle. The listed numbers are hours of RA on the celestial sphere. Using your collected RA data for each 15th or 20th, 21st, or 22nd of the month only (ignoring the 1st of the month data), plot the position of the Earth on the dashed orbit line.

Remember that the RA data you have represent where the Sun appeared to us as observers on the Earth, meaning that we were on the opposite side of the Sun. For example, if we saw the Sun at 3h RA on a particular date, that would mean that the Earth was at the position indicated by the circle below (from that position, we would see the Sun along our line of sight in front of 3h 00m RA). Mark the Earth's position with a small circle, indicate the date, and move to the next month.

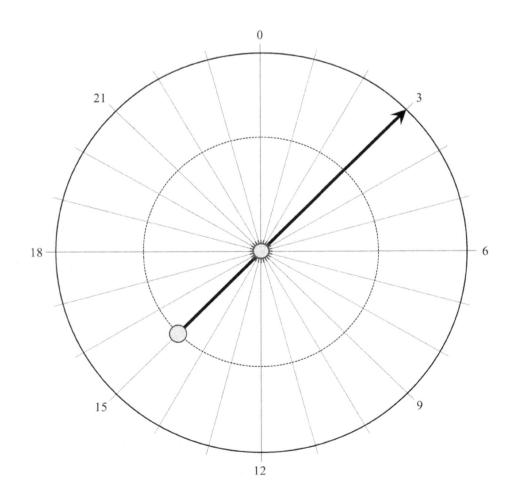

UNIT 5.2   THE LENGTH OF THE DAY

OBJECTIVE
To learn about the varying length of the daylight hours across the face of the Earth over the year

INTRODUCTION
Daylight hours begin when the Sun rises over an observer's eastern horizon, progresses to High Noon when the Sun transits across the meridian and reaches its greatest height, and end when the Sun sets in the west. Earth's day is a product of the rapid 24 hour rotation of our planet, which brings the Sun into view and out of view for different observers on the surface of the Earth.

The various seasons and weather conditions on Earth arise from the planet's axial tilt and its revolution around the Sun. Warm summer temperatures are a result of prolonged exposure to steady, direct sunlight, with and lack of solar heating in the winter leading to hemisphere-wide cooling.

PROCEDURE
Using the datasheet provided with major world cities, latitudes, and lengths of daylight hours, create a graph of sunlight versus latitude for each month. Beginning with the high northern latitude city of Anchorage, Alaska, plot daylight hours versus latitude using a letter *M* for March data points, *J* for June data points, *S* for September data points, and *D* for December data points. Move through the list and connect matching months with a smooth best-fit line.

DATASHEET #1

Location	Latitude	Day	Length of Daylight Hours
Anchorage, AK	61° N	Mar 20	12h 15m
		Jun 21	19h 22m
		Sep 22	12h 13m
		Dec 21	5h 27m
London, UK	51° N	Mar 20	12h 10m
		Jun 21	16h 38m
		Sep 22	12h 12m
		Dec 21	7h 49m
Dallas, TX	32° N	Mar 20	12h 08m
		Jun 21	14h 18m
		Sep 22	12h 07m
		Dec 21	9h 58m

(Continued)

Location	Latitude	Day	Length of Daylight Hours
Hong Kong	22° N	Mar 20	12h 06m
		Jun 21	13h 30m
		Sep 22	12h 07m
		Dec 21	10h 46m
Singapore	1° N	Mar 20	12h 06m
		Jun 21	12h 12m
		Sep 22	12h 06m
		Dec 21	12h 02m
Johannesburg, ZA[1]	26° S	Mar 20	12h 07m
		Jun 21	10h 30m
		Sep 22	12h 06m
		Dec 21	13h 46m
Santiago, Chile	33° S	Mar 20	12h 07m
		Jun 21	9h 56m
		Sep 22	12h 06m
		Dec 21	14h 22m
Christchurch, NZ	42° S	Mar 20	12h 10m
		Jun 21	9h 12m
		Sep 22	12h 06m
		Dec 21	15h 10m

[1]City located in South Africa; previous Dutch name: Zuid-Afrika

Name _____ Id _____

Due Date _____   Lab Instructor _____   Section _____

Worksheet # 1

Length of Daylight Hours

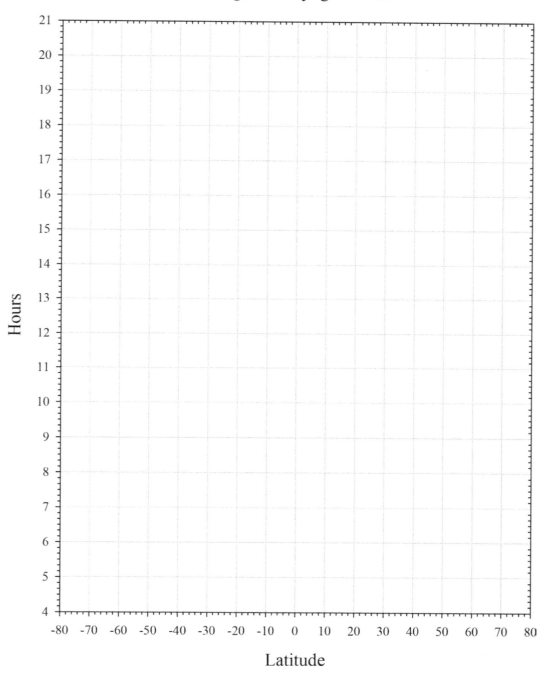

Latitude

Name _____ Id _____

Due Date _____   Lab Instructor _____   Section _____

Worksheet # 2

Answer the following post-lab questions about the length of daylight hours on Earth, considering your graphs, data, and knowledge of seasons.

1. Using your chart, what would you predict the length of a summer day to be for the Drake Passage, the southernmost tip of Chile located at 56° S latitude?

2. Referring to the question above, what month would that summer day fall on?

3. How long are the night time hours in Anchorage, Alaska on December 21st?

4. If the Earth had no axial tilt, how would the graph differ?

5. Assuming a perfect circle for Earth's orbit around the Sun (which is almost the case), how would seasons in general differ?

UNIT 5.3 EARTH'S GEOGRAPHICAL COORDINATE SYSTEM

OBJECTIVE

To become familiar with Earth's geographical coordinate system, latitude and longitude, that allows determinations of surface locations

INTRODUCTION

In everyday life the Earth appears as a flat plane to our senses, which may have led to the ancient understanding of a flat Earth. In order to designate a certain location, we usually are satisfied specifying a direction and a distance. For example, you may say that city X is 12 miles north-west of your current location. In this way, you serve as the center of your own coordinate system, with positions measured in reference to your location (this local coordinate system — called the Altitude-Azimuth system — will be further covered in Unit 5.4).

It would be difficult for different people, living at other places across our planet, to correlate their own individual coordinate systems with each other. Greek mathematicians of the Golden Age of the Hellenistic world used geometry and observations to show that the Earth was decidedly not a flat surface but instead a sphere. Much later, a coordinate system was introduced which could designate locations on the Earth's surface. The coordinates would be identical for any target, regardless of the observer's location. This coordinate system is the Latitude-Longitude system (also known as the equatorial coordinate system).

It consists of a reference plane—the equatorial plane, and its poles, the N- and S-poles. The equator plane is the fundamental *great circle*, its center is Earth's center, and its plane is perpendicular to Earth's rotational axis. At right angles to this fundamental great circle is the family of secondary great circles, the *meridians*. They also have the center of the Earth as their center, but go through both the N- and S-poles, and intersect the equator at right angles.

The geographical *latitude*, θ, of a location on the Earth's surface is the angle between the equator and this location, measured along this location's meridian. The latitude of the equator is set at 0°, with the North Pole at 90°N and the South Pole located at 90°S. All locations with the same latitude lie on a latitude circle or a parallel (these lines are parallel to the Earth's equator, hence the name). The parallels are not great circles because their center is *not* the center of the Earth. The circumference of these parallel lines of latitude become smaller and smaller as one moves toward the poles, with the equator measuring the longest path around the globe.

The geographical *longitude*, φ, of a location is the angle between a certain point on the Earth's equator with longitude 0°, and the point where the location's meridian intersects the equator (see Figure 2). The choice of the 0° longitude meridian is arbitrary. For historical reasons, the meridian which goes through the observatory in Greenwich (London), England was chosen to have 0° longitude.

Instead of specifying an angle from 0° all the way around to 360°, Longitude is expressed either from 0° to 180° East of Greenwich or from 0° to 180° West of Greenwich. The 180° meridian mostly coincides with the so-called International Dateline. All locations with the same longitude lie, of course, on the same meridian and have the same time, as we shall see later.

a) LATITUDE DETERMINATION

The angle between the horizon and the North Star Polaris is your geographical latitude. Here again we make an approximation: Strictly speaking, the angle should be between the horizon and the North Celestial Pole (NCP), Polaris is 0.8° off at present. From now on, we will use "NCP" and "Polaris" interchangeably. If you are interested where the NCP lies with respect to Polaris, consult a map of circumpolar constellations.

Latitude measures the angle between the equator (lying on the surface of the Earth) to the center of the planet, to some geographic location (also lying on the surface of the Earth). The maximum latitude in 90° at the poles (a right angle made of two tines meeting at the center of the Earth). It is standard nomenclature that positive latitudes fall above the equator and negative latitudes lie south of the equator.

b) LONGITUDE DETERMINATION

By historical treaty in 1884, the line of longitude running through the Royal Observatory in Greenwich, England, was designated as 0° longitude, the line against which all other longitudes would be measured. That line divides the world into eastern and western halves. The Prime Meridian runs from the North Pole in the Arctic Ocean down through the North Sea, travels through Britain, France, Spain, Algeria, Mali, Burkina Faso, Togo, Ghana, and through thousands of miles of the Atlantic Ocean before terminating at the South Pole, in Antarctica. On the opposite side of the globe, the Prime Meridian becomes the Antemeridian at 180° longitude (where parts of it define the International Dateline).

An observer's longitude plays a role in the reckoning of time, most obvious in the convention of time zones. The keeping of time is intimately related to the rotation of the Earth and motions of the Sun. Since the Earth rotates 360° in a 24 hour day, each measured hour is actually a 15° rotation of the planet. A simple manner of keeping time involves tracking the motion of the Sun. High Noon is the time when the Sun is directly halfway across the sky (this is the moment when the Sun's altitude is greatest, with the Sun at its highest point above the horizon). A clock set to exactly 12:00 at this moment is keeping "local solar time" (or mean solar time). Observers at different longitudes will see the Sun in different positions in their sky, even over short distances. An observer in New York (74° longitude) may see the Sun at its highest point at 12:00 PM according to their local clock. An observer in San Diego, California – 2400 miles and away at 117° longitude – will see the Sun still low on the horizon in the morning sky. The two cities – located 43° apart – are separated by 2 hours and 52 minutes. Every 1° of longitudinal separation would lead to a discrepancy of 4 minutes in terms of local solar time.

In an age of long distance communication, local time could lead to chaos in timing and schedule keeping. This was of great concern in the railroad industry, where trains would schedule times of delivery and times of intersection crossings. With so many different time keepers along the train routes, it became a matter of safety to standardize time keeping. For example, the moment of High Noon (12:00) recorded in New York City would register as 11:48 AM on a clock set to local time in Washington DC (at 77°W), 12:12 PM to a locally set clock in Boston (at 71°W), 11:36 AM in Pittsburgh (at 80°W) and 11:20 AM in

Atlanta (at 84°W). If a train leaving New York for Atlanta was scheduled to cross an intersection in Richmond, Virginia at 12:55 PM, operators may not know if the 12:55 PM was based on the time the train was keeping in NY as its starting point, or the time in its destination city in Atlanta, or the local time of the crossing point in Virginia. In order to eliminate the complexities of keeping track of local time, the concept of Standard Time was introduced in the form of time zones. All of these eastern cities, counties and towns were consolidated into Eastern Standard Time. The meridian of 75°W − which runs from the North Pole, passes through Philadelphia, and terminates at the South Pole − defines Eastern Standard Time noon. All localities within the boundaries of the Eastern Time Zone measure their clocks identically to 75°W local time. Central Standard Time is measured relative to the 90°W meridian. Mountain Standard Time approximately follows local time of the 105°W meridian, and Pacific Standard Time is approximately centered on the 120°W line of longitude. Time zones are spaced 15° and 1 hour apart.

Finally there is Universal Time (UT), which is measured as the local solar time of Greenwich, England at 0° longitude. 12:00 noon UT is the moment that the Sun is at its highest point in Greenwich (and all points on the 0° meridian) and may be thought of as a shared, worldwide time standard. The timing of most astronomical phenomena are recorded in UT for the sake of consistency for all observers across the globe. Therefore, observers in Los Angeles, California (118°W), may measure several valid times. At 12:00 noon UT, the Sun is at its highest point in Greenwich's sky and thus a Universal Time clock will read 12:00 PM. The Pacific Standard Time Zone at 120° is 8 hours behind Greenwich time, so a Standard clock will read 4:00 AM (120°/15° per hour = 8 hours difference). With Los Angeles at 118°W, that places the city 472 minutes (1° separation = 4 minutes difference, so 118° = 472 minutes, or 7 hours 52 minutes) behind Universal Time. A local solar clock, therefore, would read 4:08 AM. 12:00 PM, 4:00 AM, and 4:08 AM may all be acceptable times, depending on the type of time being measured. 5 minutes later, the UT clock will read 12:05 AM, the Standard clock will read 4:05 AM, and the local clock will read 4:13 AM.

For an observer in Seattle (122°W), their UT clock will read 12:00 PM, their Standard clock will read 4:00 AM (as they are still in the PST, defined by the 120° meridian), and their local clock would read 3:52 AM (8 hours and 8 minutes, or 488 minutes, behind Greenwich).

In summary, all observers − no matter their location on planet Earth − share the aptly named Universal Time, which is identical to the local mean solar time along the 0° meridian. The Earth is segmented into 24 standard time zones (matching the 24 hours in the Earth's day), centered on meridians of 0°, 15°E and W, 30° E and W, etc. Time zones in the eastern longitudes record local and standard times ahead of Universal Time while time zones in the western longitudes record times behind Universal Time. Generally, new time zones begin at half hour increment differences from UT. As an example, the Eastern Time Zone is 5 hours behind UT (−5 hours), and includes local solar times running from 4h 30 minutes to 5h 30 minutes behind UT. Consider the cities Cleveland, Ohio, at 81.69° and Milwaukee, Wisconsin, at a longitude of 87.91°. Cleveland is 5.45 hours (5h 27m) behind UT, placing it in the −5 hours Eastern Time Zone along with Philadelphia (5h 0m behind) and Bangor, Maine (4 hours 35 minutes behind). Milwaukee has a time difference of 5.86 hours from Greenwich, England, which rounds up to a whole number of −6 hours and 6 time zones behind Greenwich.

Observers within these time zones share a common time with one another. The continental US is divided into Eastern, Central, Mountain, and Pacific Time zones. Finally, observers along meridian lines may have highly localized mean solar times, based on the specific position of the Sun in the local sky.

ILLUSTRATIONS

Figure 1. The city of Seattle, Washington, lies at 47° north latitude. The line of 47N (shown as the parallel dashed line) creates at 47° angle to the equator. This circle runs around the globe and intersects other cities such as Munich, Germany; Budapest, Hungary; and Ulaanbaatar, Mongolia. Whether the angle is measured at the intersection point on the left side (as shown) or the right side, the angle will measure as 47°.

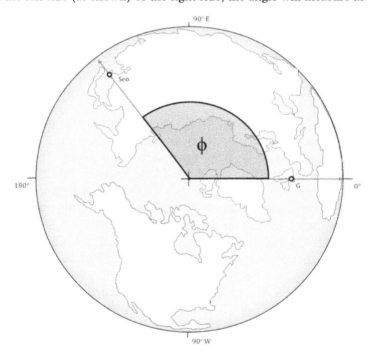

Figure 2. A top-down view of the Earth, centered on the North Pole. The city of Seoul, South Korea, lies along the 127° east longitude meridian. The angle φ is measured from the Prime Meridian (set arbitrarily at Greenwich, England). As every 15° of longitude is a difference in 1 hour, Seoul lies 8.467 hours ahead of Greenwich (which rounds up to 9 full time zones ahead of Greenwich). Other cities which lie along the 127°E longitude − Yakutsk, Russia, near the Arctic Circle or Darwin, Australia in the Southern Hemisphere − would also reckon local solar times 8hr 28m later than Greenwich. 2:30 PM UT will be measured as 11:30 PM Standard Korea time and 10:58 PM local solar time.

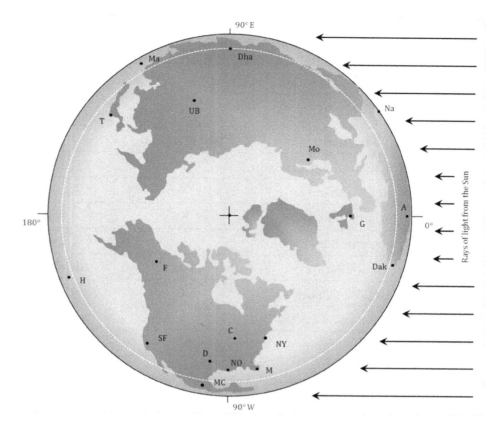

Figure A

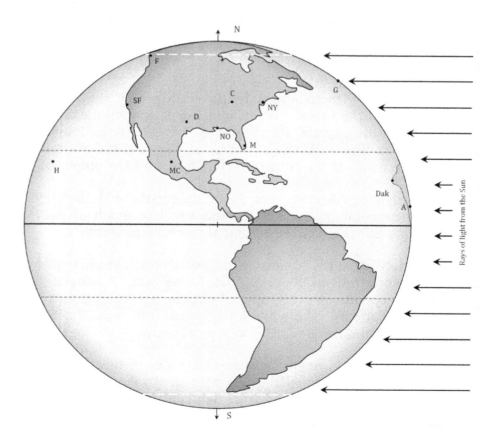

Figure B

PROCEDURE

Study Figures A and B and recognize the orientation of Earth's globe in each. Figure A shows the Earth as seen from directly above the North Pole (with the planet's axial north pole marked with a crosshair at the center of the image. Major world cities are marked with the following abbreviations:

Abbreviation	Location	Abbreviation	Location
G	Greenwich, England	H	Honolulu, Hawaii
A	Accra, Ghana	F	Fairbanks, Alaska
Dak	Dakar, Senegal	SF	San Francisco, California
Mo	Moscow, Russia	D	Dallas, Texas
Na	Nairobi, Kenya	MC	Mexico City, Mexico
Dha	Dhaka, Bangladesh	C	Chicago, Illinois
UB	Ulaanbaatar, Mongolia	M	Miami, Florida
Ma	Manila, Philippines	NY	New York, New York
T	Tokyo, Japan		

The circumference of the globe represents the planet's equator at 0° latitude. The dashed white line is the Tropic of Cancer. From this perspective, the Earth is seen to rotate counter clockwise (with the rays of the Sun coming from the right, defining the daytime and nighttime hemispheres). As oriented, the globe is shown with Greenwich, England at High Noon. The points of 0° longitude (the Prime Meridian), 90° east longitude, 180° longitude (antemeridian, with parts of this longitude line forming the International Dateline), and 90° west longitude are marked around the Earth's circumference.

If given a local time, you may use the relationship that 1 hour = 15° = 60' and 1 minute = 0.25° to determine the longitude of the city at that instant. If you are without a local time, you may measure the angle between the city and Prime Meridian to record the longitude (as was done with Seoul, South Korea, in Figure 2). Use a protractor placed at the exact center of Figure A (with the protractor's zero mark on the North Pole) to determine the longitude of the cities.

Figure B presents the Earth in profile view, centered on the equator with the Western Hemisphere in view in its entirety. The incoming light from the Sun is falling equally on the North and South Hemisphere, signaling that this is either the autumnal or vernal equinox.

The dark central horizontal line is the equator, with the tick mark at the halfway point marking the exact middle of the western hemisphere (90° west longitude, 0° latitude). The dashed gray horizontal line above the equator is the Tropic of Cancer; the dashed gray line south of the equator is the Tropic of Capricorn. The white long-dashed lines near the North Pole and the South Pole define the Arctic and Antarctic Circles, respectively.

Remember that latitude defines an angle between the center of the Earth and a point on its surface. Lines of latitude are always parallel to the equator. To determine the latitudes of cities in the Western Hemisphere, you may draw a thin, light, straight parallel line through that city, making it long enough to connect to the circumference of the Earth. Using a protractor placed at the center of the Western Hemisphere, measure the angle between the equator and the point on the circumference where your parallel line connected. That angle defines the city's latitude.

Name _____ Id _____

Due Date _____ Lab Instructor _____ Section _____

Worksheet # 1

Location	Local Solar Time			Standard Time	UT	Latitude in Degrees	Longitude in Degrees
	HR	MIN	AM or PM				
Greenwich, UK	12	00	noon	12 noon	12 noon	52° N	0°
Accra, Ghana					12 noon	4° N	0°
Dakar, Senegal	10	50	am				
New York, NY							
New Orleans, LA	6	00	am	6 am CST		30° N	
Chicago, IL							87° 42' W
Dallas, TX							97° W
San Francisco, CA	3	50	am				
Fairbanks, AK							147° W
Honolulu, HI	1	28	am				
Tokyo, Japan						35° N	
Dhaka, Bangladesh	6	00	pm		12 noon		
Moscow, Russia	2	29	pm				

What is the latitude of the equator?

What is the latitude of the Tropic of Cancer?

What is the latitude of the Tropic of Capricorn?

What is the latitude of the Arctic Circle?

UNIT 5.4 THE ALTITUDE AND AZIMUTH OF THE SUN

Objective
To learn how the seasons arise from the changing path of the Sun through the sky

Introduction
Earth owes its four major seasons to the planet's axial tilt and its path around the Sun. The planet Earth is tilted 23.5° on its spin axis, which – to an outside observer – would cause the planet to appear bowed over with its North Pole oriented toward the star Polaris. As the Earth orbits the Sun, its pole remains oriented toward Polaris. During the northern hemisphere's summer months, the Earth's northern hemisphere is bowed toward the Sun, causing the northern half of the planet to be subject to extended hours of direct sunlight. On the other side of its orbit (where the Earth winds up six months later), the northern hemisphere is now tilted back away from the Sun, entailing shorter daylight hours and a Sun that is lower in the sky. Figure 1 and 2 demonstrate the orientation of the Earth at different points in orbit.

Day by day, as the Earth winds around in orbit, the Sun's path through the sky will change from the previous day. Moving from summer through autumn and into winter, the Sun's maximum height in the sky will decrease bit by bit until bottoming out on the first day of winter (called the winter solstice). After that day, the descent of the Sun will reverse course and – every day until the summer solstice – the Sun's maximum height in the sky will increase day by day. "Solstice" translates as "Sun standing still" and refers to the reversal of the Sun's changing daily motion. The equinoxes occur at points in orbit where neither the North Pole nor the South Pole points toward the Sun, leading to days which are equal-length daylight and darkness and a Sun that rises exactly due east and sets exactly due west.

To accurately track the Sun's motion through the sky, a local coordinate system is needed. In this case, we use the altitude and azimuth system. Altitudes measure an angular distance above or below the horizon. 0° altitude designates an object sitting right on Earth's horizon, either on the verge or rising in the east or setting in the west. 90° altitude is called the *zenith*, the point directly above an observer's head. −90° altitude is called the *nadir* or the point directly below an observer. The second part of this local coordinate system is called azimuth, and refers to an angular direction along the horizon. The familiar compass coordinates north, south, east, and west are four directions corresponding to azimuth directions. The azimuth system starts at 0° (which is compass direction due north) and winds to 90° (representing exact east). Continuing along the horizon, the azimuth angle increases to 180°, representing due south, exactly opposite from north. Proceeding from there, 270° marks exact west and azimuth increases until reaching 360° (which rolls back to 0°). In the northern hemisphere, the Sun reaches its highest point in the sky at High Noon, which is the precise moment that the Sun reaches an azimuth of 180° and transitions from rising to setting.

Tracking the altitude of the Sun at High Noon as well as the azimuth of the rising or setting Sun, drastically different positions of the Sun and the rise of the seasons can be demonstrated. Clearly, the hot days of summer are a result of the Sun shining down on the northern hemisphere at a direct angle, with the Sun remaining in the sky for an appreciably longer time than it does in the winter time.

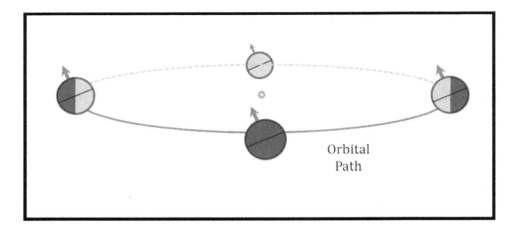

Figure 1. While not drawn to scale, this schematic shows the various positions of the Earth at points in its orbit marking the northern hemisphere's winter solstice (leftmost position), vernal equinox (foreground Earth), summer solstice (rightmost position) and autumnal equinox (background Earth). Notice that the North Pole (arrow) always points in the same direction. The darkened semi-circle shows which portion of the Earth is facing away from the Sun (representing the night-time side of the planet)

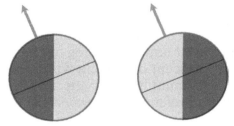

Figure 2. Simple schematic of the light distribution across the planet during northern hemisphere summer (right side) and winter (left side). The arrow represents the North Pole with the equator cutting the planet into northern and southern hemispheres. Looking at the Earth like a pie chart, notice that the summer-oriented Earth (on the right) has its northern hemisphere receiving a majority of the sunlight (~55% of the northern hemisphere is in light, as opposed to the 45% lit southern hemisphere). This extra sunlight leads to longer, hotter days with the Sun appearing to reach a higher maximum altitude in the sky.

PROCEDURE

For the dates given on Worksheet #1, record the altitude of the Sun at High Noon (when the Sun reaches its greatest angular height above the horizon as it crosses the meridian at an azimuth of 180°) as well as the azimuth of either the rising or setting Sun (or both).

On Worksheet #2, plot the data points for the altitude versus date for the given months. Connect the data points with a smooth best-fit line. Depending on whether you have viewed sunrises, sunsets, or both, plot your azimuth readings versus dates on the given graphs. For Worksheet #3, label the area above 270° as north-west and the area below 270° as south-west. Likewise, for Worksheet #4, label the area above 90° as south-east and the area below 90° as north-east. Connect the data points with a smooth best-fit line to show the gradually changing position of the Sun along the eastern or western horizon.

Using your graphs, data, and the information in the lab, answer the post lab questions dealing with the seasons and positions of the Sun.

Name _____ Id _____

Due Date _____ Lab Instructor _____ Section _____

Worksheet # 1

Altitude Angular height of the Sun above the horizon at High Noon	
Date	**Altitude** **(in degrees)**
Mar 20	
Apr 21	
May 21	
June 21	
July 21	
Aug 21	
Sept 22	
Oct 21	
Nov 21	
Dec 21	
Jan 21	
Feb 21	
Mar 20	

Azimuth Angular direction of the Sun's setting or rising position with respect to due north	
Date	**Azimuth** **(in degrees)**
Mar 20	
Apr 21	
May 21	
June 21	
July 21	
Aug 21	
Sept 22	
Oct 21	
Nov 21	
Dec 21	
Jan 21	
Feb 21	
Mar 20	

Name _____ Id _____

Due Date _____ Lab Instructor _____ Section _____

Worksheet # 2

Altitude of the High Noon Sun

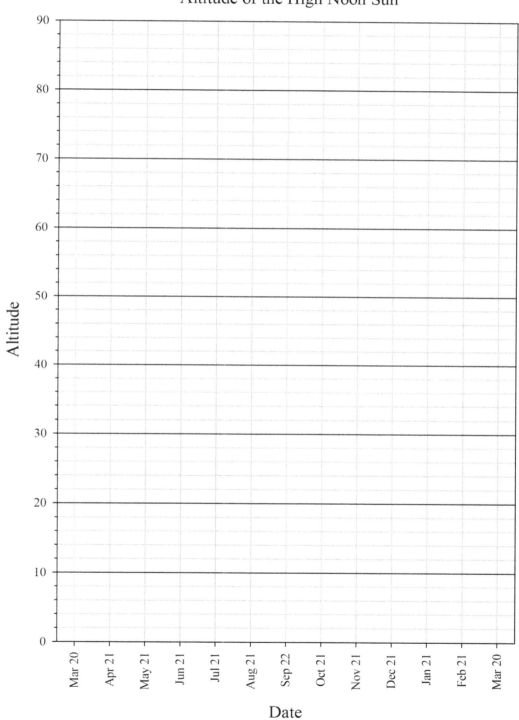

Date

Name _____ Id _____

Due Date _____ Lab Instructor _____ Section _____

Worksheet # 3
Azimuth of the Setting Sun

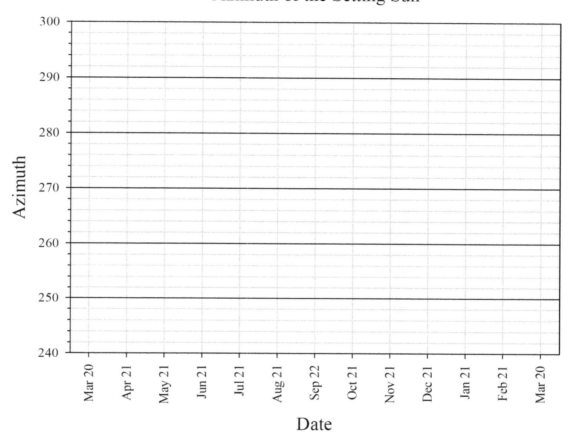

Name _____ Id _____

Due Date _____   Lab Instructor _____   Section _____

Worksheet # 4
Azimuth of the Rising Sun

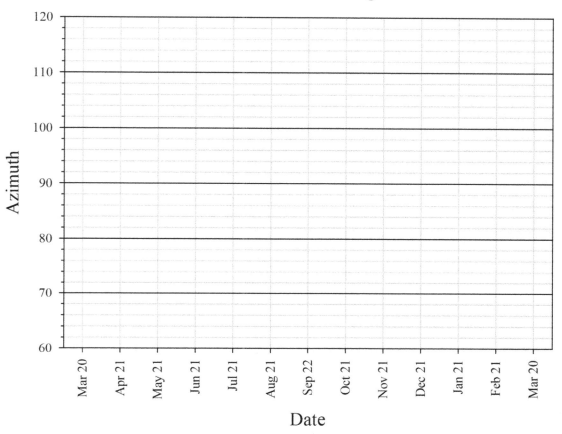

Name _____ Id _____

Due Date _____   Lab Instructor _____   Section _____

Worksheet # 5

Using your data and graphs, along with your knowledge of astronomy, answer the following questions:

1. On which date or dates of the year does the Sun set exactly due west?

2. On which date or dates of the year does the Sun set farthest to the north?

3. On which date or dates of the year does the Sun set farthest to the south?

4. According to your altitudes, when during the year is the Sun "directly overhead" as seen from your location? ("Directly overhead" meaning at the zenith point, or 90° altitude)

5. What would you estimate the *altitude* of the Sun to be – in degrees – at *sunset* on July 9th?

6. You have recorded the azimuth of the setting Sun for March 20th, but what was the azimuth of the Sun at sunrise on that day?

7. Explain what is meant by "High Noon". When does it occur?

8. Archeologists find an ancient monument composed of two stone slabs spaced closely together and angled toward the horizon, facing 253°, so that the setting Sun could be seen between the monuments only on specific days. They were probably used to keep track of significant holidays. Using your graphs, determine on what days the Sun will set between these monuments.

UNIT 5.5   THE CELESTIAL SPHERE

OBJECTIVE

To become familiar with the celestial sphere, including the astronomical coordinate system based on Right Ascension and Declination, which allow locating celestial objects

INTRODUCTION

The Right Ascension (RA) and Declination (Dec) system is the astronomical equivalent to the geographical *latitude* and *longitude* system. The longitude is measured along the Earth's equator from an arbitrary point, the Greenwich meridian. The RA is measured along the celestial equator from an arbitrary point, the *vernal equinox* (VE). Earth latitude is measured as an angle from the Earth's equator to the location in question, along the location's meridian. Declination is measured as the angle from the celestial equator to the location in question, along the location's hour angle (HA).

The fundamental great circle of the RA and Dec system is the celestial equator with its poles, the North Celestial Pole (NCP) and South Celestial Pole (SCP). The center of the circle is the center of the Earth.

While the choice of Earth's 0° meridian was arbitrary from a scientific point of view, the choice of 0° or 0 hours RA is not wholly arbitrary. The Earth spins on its axis which is fixed in direction relative to the distant stellar background (neglecting precession, the slow wobble of the Earth's orientation as well as extremely-long-term changes of stellar positions).The celestial equator, always at right angles to the axis of the Earth, is therefore also fixed in space. At the same time the Earth revolves around the Sun in a plane which is inclined to the equatorial plane. Therefore, there are two points where the equatorial plane intersects the plane of the ecliptic. These two points are the *equinoxes*. One of them, the vernal equinox, is the origin of the RA scale.

The angle of inclination between the equatorial plane and the ecliptic plane is the so-called *obliquity of the ecliptic*. Its value is 23.5°. Another, equally valid, definition of the ecliptic is: the ecliptic is the apparent path − as seen from Earth − of the Sun relative to the stellar background. It is centered on the belt of the 12 Zodiac constellations. Strictly speaking, the ecliptic crosses the boundary of 13 constellations.

The relationship between all of the quantities described above is shown in Figure 1.

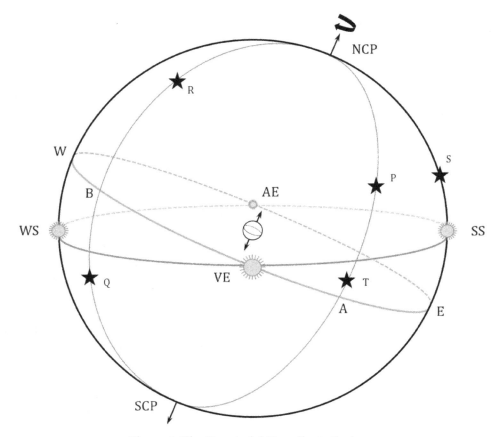

Figure 1. The Equatorial Coordinate System.

The following abbreviations are used in Figure 1:

NCP	North Celestial Pole	VE	Vernal Equinox
SCP	South Celestial Pole	AE	Autumnal Equinox
SS	Summer Solstice	E	East
WS	Winter Solstice	W	West

The Earth rotates counterclockwise in the figure above, moving from W toward E (Earth rotates from west to east). The celestial sphere seems to rotate in the opposite direction from east to west, as indicated by the small arrow on top of the NCP. This leads to the stars, Sun, Moon, and planets appearing to rise in the east and set in the west. The position of the Sun is shown for both equinoxes and solstices on the ecliptic, indicated by points VE, SS, AE, and WS.

Study Fig. 1: Note the Earth's N and S poles and the NCP and SCP at the intersection of the extension of Earth's axis with the celestial sphere. The tilted ellipse (tilted at an angle of 23.5°) is the celestial equator, the extension of Earth's equatorial plane out into space. The two points of intersection between the equatorial and the ecliptic plane are the equinoxes, with the vernal equinox in front and the autumnal equinox below the plane of the paper. The RA of an object is measured along the celestial equator, starting with 0° or 0 hr at the VE (vernal equinox) in an easterly (counterclockwise) direction. It is expressed either in degrees, from 0° to 360°, or in hours (sidereal) from 0 hr to 24 hr.

The DEC of the celestial equator or of any object on the celestial equator is 0°. The DEC of the NCP is +90°, that of the SCP −90°. + or − are used instead of N and S.

PROCEDURE

Explore the following examples:

1. Consider the Sun to be at point WS on the ecliptic. Earth's southern hemisphere would receive more light than the northern one, the S pole is in sunlight, the N pole in shade. It is winter on the northern hemisphere, the Sun is in the Winter Solstice (WS), the date would be December 21. What are the Sun's coordinates?

   Answer: The Sun's hour circle is NCP-W-WS-SCP. It intersects the celestial equator at point W. The RA of the Sun is then the arc VE-E-AE-W corresponding to 270° or 18 hr. The DEC of the Sun is the arc W-WS or −23.5° (it is negative (−) because it is located below the celestial equator).

   During the next quarter year the Earth revolves one quarter the circumference around the Sun or the Sun seems to move one quarter around the ecliptic from point WS to VE. It is then March 21, the vernal equinox. The coordinates of the VE are: RA = 0° or 0 hr; DEC = 0°, because the VE is on the celestial equator.

2. What are the coordinates of star P? Answer: Star P's hour circle is NCP-P-A-SCP. It intersects the celestial equator at point A. The RA of star P is therefore the arc VE-A, or about 30° or 2 hr as judged from the drawing. The DEC of star P is the arc A-P (along P's hour circle). As judged from the drawing, it is about 50° (it is positive (+) because it is located above the celestial equator).

Name _____ Id _____

Due Date _____   Lab Instructor _____   Section _____

Worksheet # 1

From Figure 1 read the coordinates, RA, and Dec. Express them in arc segments (e.g., the arc segment VE-A). Give an approximate value of the angles in hours and/or degrees

Object	Arc	RA Estimated Value		Declination Estimated Value	
		in degrees	in hours	in degrees	in arc
Autumnal Equinox (AE)					
Summer Solstice (SS)					
Star Q					
Star R					
Star S					
Star T					
South Celestial Pole (SCP)					
Point E					

UNIT 5.6   PHASES OF THE MOON

OBJECTIVE
To demonstrate how the changing phase of the Moon, its rising time, and its setting time are linked to its position in orbit and angular distance from the Sun

INTRODUCTION
The continually changing phases of the Moon served as one of the first and most reliable calendars for ancient people. The Moon predictably progresses from its new phase, becoming more illuminated as it advances through its fully lit full phase before whittling back down to a new Moon. The cause of the familiar lunar phases – and their timing – is an effect of the changing Sun-Earth-Moon angle. Like the Sun and the major planets, the Moon moves against the background stars, passing through the stars of the ecliptic as it completes its orbit about the Earth. Using the celestial sphere and the right ascension of the Sun and Moon, it is possible to track exactly where the Moon is in its orbit around the Earth, as the lunar phase and separation from the Sun are related.

The Earth rotates at high speed, causing the Moon, stars, planets, and Sun to rise, arc through the sky, and set. We define our day by the position of the Sun: sunrise, High Noon, sunset, and astronomical midnight (when the Sun is located on the side of the Earth exactly opposite to the observer). The Sun and Moon follow very similar paths through the sky; both pass through the same constellations of the zodiac, following the ecliptic. All phases of the Moon are determined by the angle between the Earth, Moon, and Sun (as shown in the illustrations in this summary).

0° separation is the (unlit) new Moon, with a 90° separation called the first quarter (where the Moon is 50% lit on the right side). Between 0° and 90° are phases called waxing crescents (less than 50% lit). 180° separation (where the Sun and Moon are on opposite sides of the sky) leads to a fully 100% lit full Moon. Between 90° and 180° are the waxing gibbous phases, displaying more than 50% illumination. The Moon reaches its last quarter (or third quarter, as the Moon is three-quarters of its way through its cycle of phases) at 270°. Waning gibbous phases define the Moon between 180° and 270°. Finally, before reaching the end of its phases and concluding the lunar month by returning to the new Moon, the waning crescent phases occur between

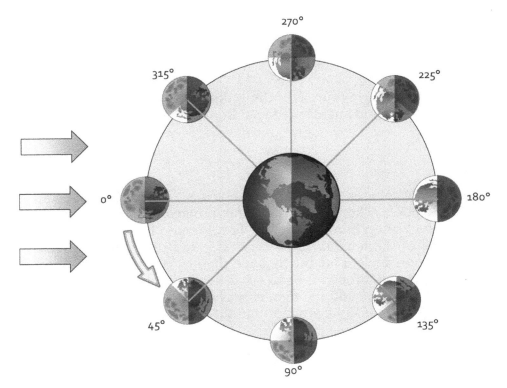

Figure 1. An illustration of the position of the Moon at select phases with Sun-Earth-Moon angles labeled. The yellow arrows indicate incoming sunlight. The gray arrow illustrates the direction of the Moon's travel through orbit (counter-clockwise revolution, the same direction that the Earth rotates). The red hemisphere is the "far side" of the Moon, which – during the Moon's orbital motion – always remains pointing directly away from the Earth. The black hemisphere is the unlit portion of the Moon which always points directly away from the Sun, since sunlight is the source of the Moon's illumination. White slices of the Moon's surface represent the portion of the Moon lit and visible from the surface of the Earth. (Sizes and distances not to scale)

270° and 360°/0°. See Figure 1 for a diagram of the Moon's phases versus angular distance from the Sun.

As stated before, 0° is the new Moon position followed by waxing crescent phases; 90° is the first quarter, followed by the waxing gibbous phases; 180° is the full Moon, followed by the waning gibbous phases; and 270° is the last quarter, followed by waning crescent phases

EQUATIONS AND CONSTANTS

$$1 \text{ hour} = 15°$$

$$1 \text{ minute} = 0.25°$$

$$1° = 4 \text{ minutes}$$

DATASHEET #1

This graph shows the Moon's separation from the Sun (i.e., its position in orbit) versus the illumination of its near side. The separation is measured in degrees, running from 0° to 360°. The illumination is expressed as a percentage.

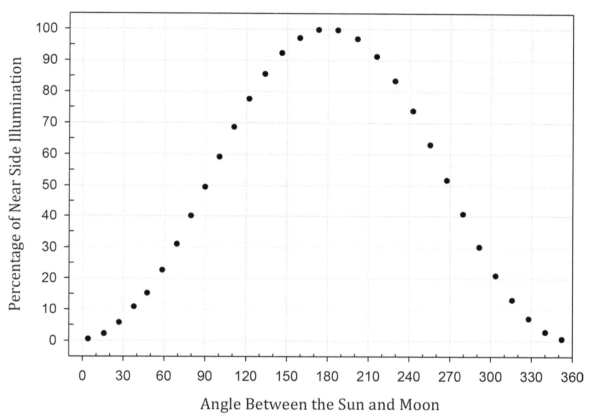

Moon Illumination Vs. Orbital Position

Name _____ Id _____

Due Date _____ Lab Instructor _____ Section _____

Worksheet # 1

Part I: Math with Angles and Time

1. Convert the following from hours into degrees:

1 hour	

6 hours	

15.5 hours	

21.33 hours	

12 hours 15 minutes	

3 hours 10 minutes	

2. Convert the following from degrees into hours:

60 degrees	

15.5 degrees	

270 degrees	

300 degrees	

23.5 degrees	

37.25 degrees	

Name _____ Id _____

Due Date _____ Lab Instructor _____ Section _____

Worksheet # 2

Part II: Phases and Position

Using the information given, your knowledge of Moon phases, and your graph of the Moon Illumination versus Position, answer the following questions:

2.1   How many degrees and hours of Right Ascension separate the first quarter Moon and the Sun?

Degrees	Hours

2.2   How many degrees and hours of Right Ascension separate the last quarter Moon and the Sun?

Degrees	Hours

2.3   How many hours of time separate the rising of the Sun and the rising of the last quarter Moon?

2.4   How many hours of Right Ascension separate the 88% waning gibbous Moon and the Sun?

2.5   Which side of the Moon is lit during the waxing crescent phase (left hemisphere or right hemisphere)?

2.6   From the surface of the Earth, does the waxing crescent Moon appear to the left or right of the Sun when they are in the sky together?

Name _____ Id _____

Due Date _____   Lab Instructor _____ Section _____

Worksheet # 3

Part III: Phases and Time

For the following questions, remember the correlation between time (in hours and minutes) and degrees. Angle separations and time separations are closely linked in astronomy and can be converted, added, and subtracted.

1. The Moon rises at 7:37 AM. At what time is the Moon 30° above the horizon?

2. The Sun rises at 5:57 AM one summer morning. At High Noon, the Sun is at the half-way point of its path through the sky and has moved 111° from its rising position. What time is High Noon that day?

3. The Sun rises at 6:12 AM while the Moon is in the first quarter phase. What time does the Moon rise that day?

4. Will a very thin waxing crescent Moon set shortly before the Sun or after the Sun?

5. Will a 33% waxing crescent rise later or earlier than a 25% waxing crescent?

6. If the Sun rises at 6:00 AM one morning while the Moon is in the full phase, what time will you see the full Moon rise that day?

7. The Moon rises at 10 AM, 2 hours and 20 minutes *after* the Sun. What phase is the Moon in and (using your graph from Datasheet #1) what percentage of the Moon is illuminated?

Name _____ Id _____

Due Date _____ Lab Instructor _____ Section _____

Worksheet # 4

Part IV: The Locations of the Moon

Below is an unscaled illustration of the Sun (to the left), the Earth (center), and the Moon's orbit, with major separations of 0°, 90°, 180°, 270° marked. Similar to the lab summary and the presentation, the Moon's motion is counterclockwise. With the given data you will be able to determine all pertinent information, such as time or angle separations between the Sun and Moon. Using a protractor and your knowledge of the Earth-Moon-Sun separations and lunar positions, mark — with a small circle — the position of the Moon being referenced in the question:

1. Mark the position of the Moon in orbit when it is 76% lit and waning.

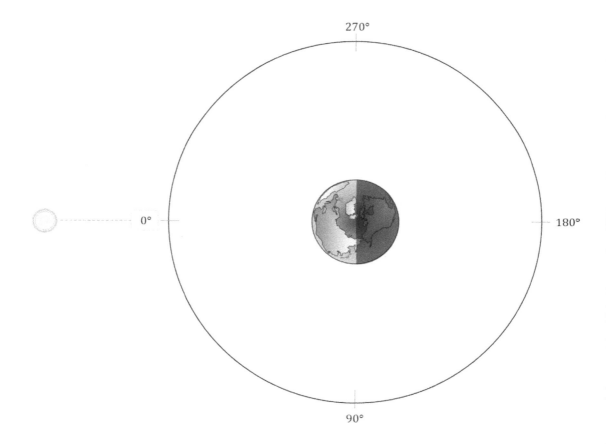

Continue....

2. Mark the position of the Moon in orbit when it sets 11 hours and 30 minutes after the Sun sets.

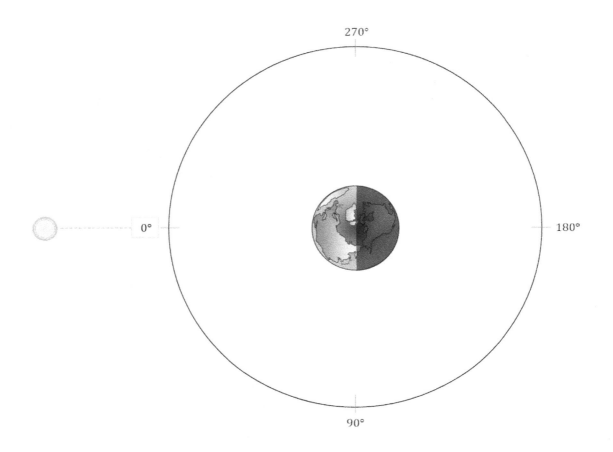

Continue....

Name _____ Id _____

Due Date _____   Lab Instructor _____   Section _____

Worksheet # 4, cont'd

3. Mark the position of the Moon in orbit when it is 25% lit and rises shortly before the Sun rises

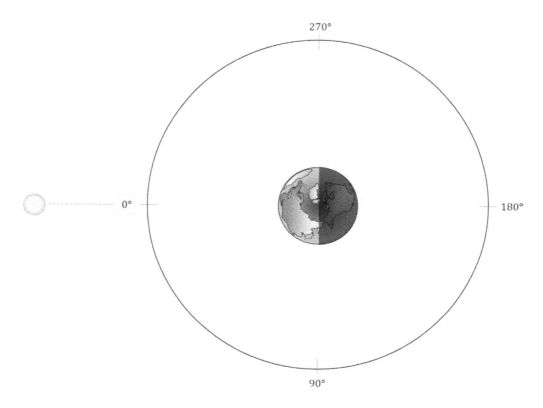

UNIT 5.7   THE APPARENT PATH OF THE MOON

OBJECTIVE

To learn how the Moon progresses through the ecliptic. This lab covers the motions of the Moon against the background stars and the changing phase

INTRODUCTION

The Moon displays visibly different phases as the month wears on, progressing from a 0% illuminated new Moon, progressing through waxing crescent phases to the 50% illuminated first quarter, transitioning into the waxing gibbous phases as the amount of light falling on the near side of the Moon increases day by day until reaching a maximum at 100% illumination for the full Moon. From there, the Moon's nearside will transition back into darkness, passing through the waning gibbous phases, reaching the last quarter at 50% illumination along the Moon's left half, progressing through the waning crescent phases and returning to the new Moon phase. This time period is referred to as the *synodic period*, and is based on the Sun-Earth-Moon angle which give rise to the various phases.

In addition to visibly changing every day, the Moon sweeps rapidly through the ecliptic, much in the same way as the Sun and planets. The sidereal period of the Moon – the time it takes the Moon to complete one 360° orbit around the Earth – can be determined from the Moon's Right Ascension. When the Moon returned to a particular point on the celestial sphere, it has completed on full circuit around the Earth. This time period is referred to as the *sidereal period*. Because of the revolution of the Earth around the Sun, the sidereal and synodic period are not equal. Since the synodic period is dependent on alignments between the Earth, Moon, and Sun (and because the Earth's revolution is continually altering the angles between these three bodies), the synodic period tends to lag behind the sidereal period. Progressing from one new Moon phase to the following new Moon phase requires that the Moon travel further than 360° around the Earth.

PROCEDURE

On Worksheet #1, collect data on the position of the Moon using a source such as Stellarium or other planetarium software. You may be assigned a starting date or you may choose one yourself. Generally, in the latter case, it is easiest to begin your observations on one of the four unique phases of the Moon, either 0% illuminated new Moon, 50% illuminated first or last quarter, or 100% illuminated full Moon. Progress forward 1 day for each data point and record the Moon's RA (in hours and minutes), its Declination (in degrees) and its illumination (a percentage if available or a phase, such as new, waxing crescent, first quarter, waxing gibbous, full, waning gibbous, last quarter, or waning crescent).

Using your data points, plot the path of the Moon on the star charts provided. Each tick mark on the declination axis is equal to 1°. Each small tick mark between the labeled hours is equal to 1/10 of one hour, or 6 minutes. For example, the labeled 12H tick mark represents 12h 0m. The first short tick mark is 12h 6m, followed by 12h 12m for the second tick mark, 12h 16m for the third and so on. When the data points have been plotted, connect the points with a smooth best-fit line.

Label all instances of new Moon, full Moon, and quarters. Also label the *regions* of waning, waxing, crescent, and gibbous (mark along the best-fit line; it isn't necessary to label every individual point as waxing crescent, waning crescent, waxing gibbous, waning gibbous, etc.)

Name _____ Id _____

Due Date _____ Lab Instructor _____ Section _____

Worksheet # 1

Data Point	Date	RA of the Moon (in hr)	Declination of the Moon (in deg)	Illumination
1				
2				
3				
4				
5				
6				
7				
8				
9				
10				
11				
12				
13				
14				
15				
16				
17				
18				
19				
20				
21				
22				
23				
24				
25				
26				
27				
28				
29				
30				
31				

Name _____ Id _____

Due Date _____   Lab Instructor _____   Section _____

Worksheet # 2

1. Over the course of the month, the Moon's RA changes steadily. Recalling that the Moon has completed one orbit when its position repeats a previous Right Ascension, how long would you estimate the sidereal period to be according to your data? What data points and dates led to this conclusion?

2. Over the course of the month, the Moon's illumination changes steadily. Recalling that the Moon has realigned with the Sun and Earth when it repeats a previous phase illumination, how long would you estimate the synodic period to be according to your data? What data points and dates led to this conclusion?

3. Does the Moon orbit the Earth with a constant speed? To show if it does or does not, use a ruler and measure a few distances between consecutive points. The longer the distance between the points, the faster the Moon has traveled. Is there any time when the Moon appears to be moving fastest? Is there a place where the Moon appears to move at its slowest pace?

Name _____ Id _____

Due Date _____   Lab Instructor _____   Section _____

Worksheet # 3

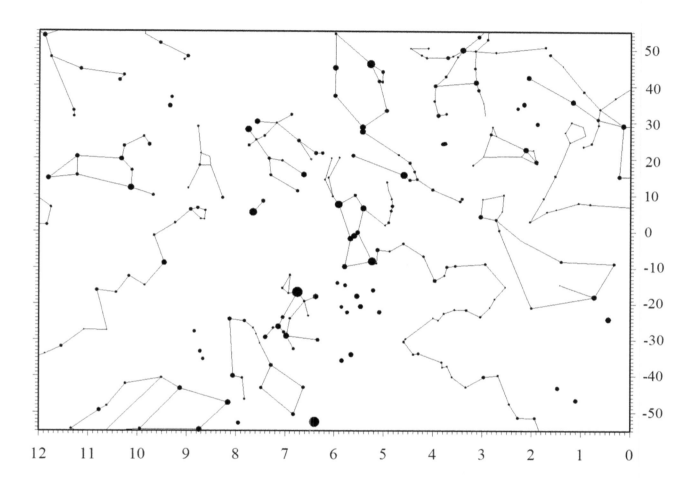

Name _____ Id _____

Due Date _____ Lab Instructor _____ Section _____

Worksheet # 4

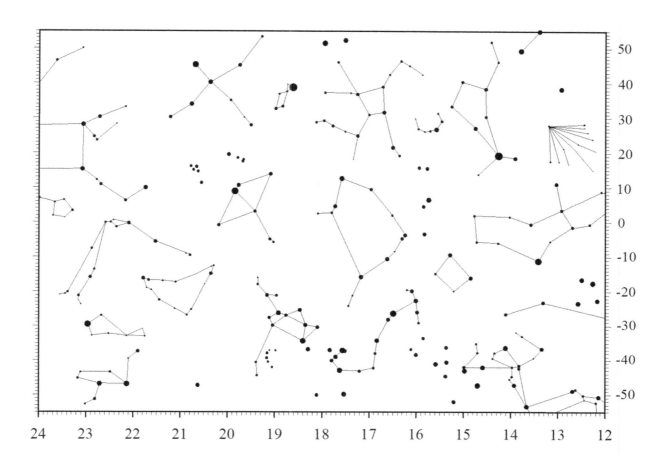

UNIT 5.8   MOON AND MARS LANDING SITES

OBJECTIVE

To identify the locations of the manned landing sites on the Moon and the landing spots of landers and rovers on Mars

INTRODUCTION

Given the daunting distances within the Solar System, much of astronomy must be done from the surface of the Earth. The closest star would require well over 25,000 years to reach due to fundamental physical limitations and the constraints of modern-day rocket technology. However, some objects are close enough and generally hospitable enough that manned and robotic missions could be sent to study these distant and exotic worlds. In particular, widely successful missions have been conducted on the Moon and Mars, utilizing both human and robotic explorers.

From July 1969 through December 1972, the space race with the Soviet Union culminated in 6 successful NASA lunar landings lasting a total of 80 hours on the lunar surface. The Apollo missions brought back to Earth hundreds of pounds of Moon rock for study, material which would turn out to be instrumental in solidifying the theory that the Moon was created by a catastrophic collision between the Earth and a now-destroyed planetary object in the early years of the Solar System (although some open questions still remain). As a consequence of the space race, a plethora of new computer, rocket, and electronic technology was created, ushering in a technological revolution.

While humans have yet to set foot on another planet, a series of successful NASA landers and rovers have dug through the soil of Mars in search for clues about a possibly watery past. Of all the planets in our Solar System, Mars seems to be the best candidate for ancient (or still present?) life. Though it is dry, cold, and with a very thin atmosphere, there is mounting evidence that Mars may have once hosted rivers, lakes, or even shallow seas of liquid water, one of the main ingredients for life.

Both of these worlds have been heavily scrutinized by teams of scientists with orbiters, landers, and rovers, with fascinating geological features – caverns, caves, extinct volcanoes, etc. – mapped and catalogued. Exceptionally detailed maps of both the Moon and Mars have been created in the past decade by missions such as the Lunar Reconnaissance Orbiter and the Mars Reconnaissance Orbiter. Just like a map of Earth, the Moon and Mars are sectioned off into latitude and longitude. For the Moon map, the prime meridian – separating the eastern hemisphere of the Moon from the western hemisphere – runs along the line of 0° longitude directly down the center of the Earth-facing side of the Moon, stretching around the Moon from pole to pole. Negative values to the left of 0° are located in the western lunar hemisphere, with positive values indicating sites east of the meridian. The Moon's equator represents 0° latitude, with positive latitudes stretching toward the north lunar pole and negative values indicating a position south of the equator. On Mars, 0° longitude was arbitrarily set by observers studying the planet in the 1800s who set the easily spotted crater *Davies* as the prime meridian of the planet. Like in case of Earth and the Moon, 0° latitude marks Mars's equator.

PROCEDURE

Included in the lab are three data tables: one containing the coordinates of the Apollo landing sites on the Moon, one with the robotic landing sites on Mars, and a table of interesting points on Mars. Using the coordinates, a ruler and the lunar and Martian maps, plot the position of the objects on the maps using a small but visible dot. Label the point with a small symbol, given in the data table (i.e., mark the point of the Apollo 11 landing site with a small "**11**")

References:

Map of Mars: http://mola.gsfc.nasa.gov/images.html

Map of the Moon: http://astropedia.astrogeology.usgs.gov/download/Moon/LRO/LROC_WAC/thumbs/Moon_LRO_LROC-WAC_Mosaic_global_1024.jpg

Datasheet #1

Mission	Lunar Latitude	Lunar Longitude	Symbol
Apollo 11	0.7° N	23.5° E	11
Apollo 12	3.0° S	23.4° W	12
Apollo 14	3.6° S	17.5° W	14
Apollo 15	26.1° N	3.6° E	15
Apollo 16	9.0° S	15.5° E	16
Apollo 17	20.2° N	30.8° E	17

Mission	Martian Latitude	Martian Longitude	Symbol
Viking 1	22.5° N	310.0° E	V1
Viking 2	48.0° N	134.3° E	V2
Pathfinder	19.1° N	326.8° E	PF
Spirit	14.6° S	175.5° E	S
Opportunity	1.9° S	354.5° E	O
Phoenix	68.2° N	234.3° E	PX
Curiosity	4.6° S	137.4° E	C

Interesting Sites	Martian Latitude	Martian Longitude	Symbol
Olympus Mons	19° N	225° E	OM
"Face" on Mars	40.7° N	9.5° E	F
Hellas Planitia	48° S	326.8° E	HP
Garni Crater	11.3° S	69.7° E	GC

- Olympus Mons is the Solar System's highest volcano currently known (which is unlikely to change), at nearly 3× the height of Mt. Everest and 2× the height of Mauna Loa / Mauna Kea, Hawaii, if measured from their deep-ocean base.
- Hellas Planitia is the third largest visible crater in the Solar System
- Garni Crater was discovered to host seasonal outflows of liquid water, which orbiting probes have seen run down the crater's slopes

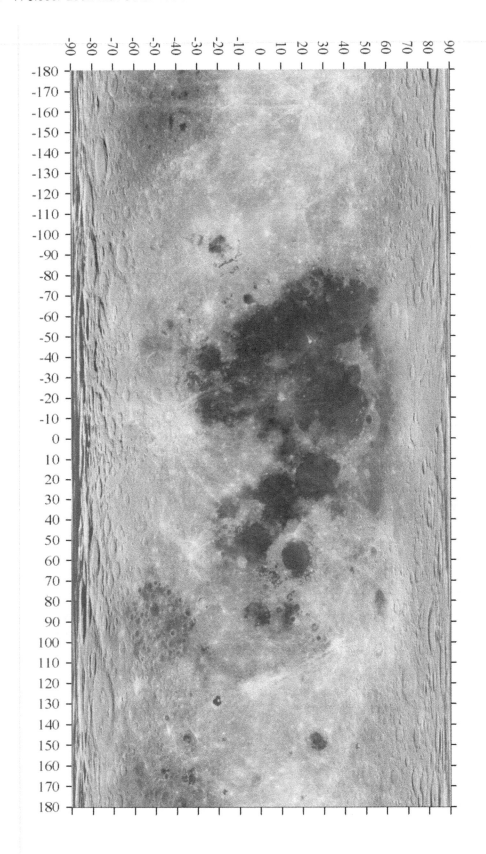

Source: NASA

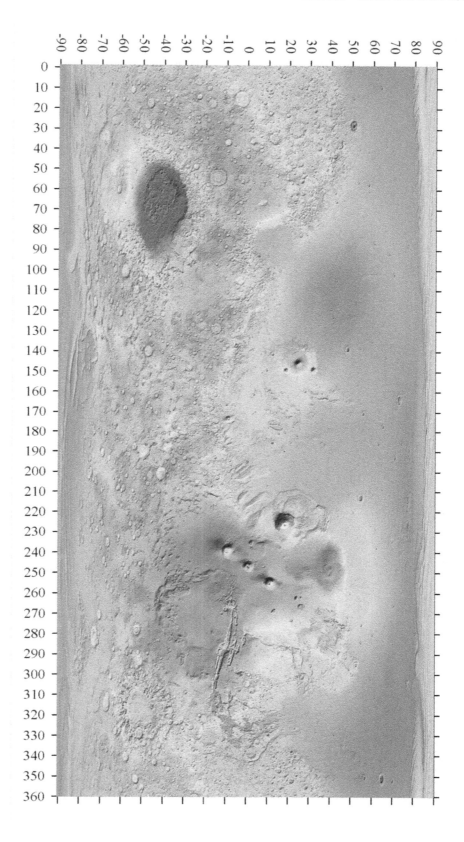

Source: NASA

Name _____ Id _____

Due Date _____   Lab Instructor _____ Section _____

Worksheet # 1

1. Which of the Apollo missions landed in the lunar highlands, the rocky, light colored terrain in the eastern hemisphere of the Moon?

2. The lunar Apennine Mountains are located high in the northern hemisphere of the Moon, between the "Sea of Serenity" (Mare Serenitatis) to the east and the "Sea of Rains" (Mare Imbrium) to the west. Which Apollo mission landed in this northern mountain range?

3. The Sea of Tranquility (Mare Tranquillitatus) is located just south of the Sea of Serenity. Which Apollo mission landed in the Sea of Tranquility?

4. Mars's axial tilt, which is 25° inclination, is slightly greater than that of Earth's, identified as 23.5° inclination. Thus, Mars undergoes the same seasonal changes – moderate equinoxes and extreme weather solstices – that Earth does, including cold, dark nights at the poles. Which of the robotic landers was most likely to be frozen and rendered inoperable by Martian winter and what leads to this conclusion?

5. Which robotic mission on the Martian surface is closest to a site where scientists found evidence that there may be seasonal, flowing liquid water?

UNIT 6: THE SOLAR SYSTEM AND BEYOND

UNIT 6.1   A SCALE MODEL OF THE EARTH

OBJECTIVE

To learn about the structure of the Earth's interior and the layers of its atmosphere

INTRODUCTION

The Earth is the Solar System's largest terrestrial planet, and – like the Solar System's other terrestrial planets – is marked by a dense rocky-metallic composition. Early in the Earth's history, the planet was in a liquefied state, with energy from impacts, compression and friction, keeping it from solidifying. Since dense liquids sink and light-weight materials buoy to the surface, the Earth became stratified, or separated into layers dependent on density. Heavy metals – iron, gold, and uranium – sunk to the center of the planet while light materials – silicon, oxygen, and aluminum – bubbled to the surface. Today, the Earth's interior is separated into various distinct sections, which geophysicists have been able to probe by carefully studying the way seismic waves originating from earthquakes propagate through the planet's interior.

The Earth's inner core is a high temperature, high density sphere of mostly iron and nickel. With a temperature coincidently comparable to that of the surface of the Sun (5780 Kelvin), it might be expected that the core should melt into a liquid. However, the high pressure exerted by the weight of the planet's overlying layers act upon the core from all directions. This pressure crushes the core into a solid crystal state. Above the inner core lies the liquefied outer core, an internal "ocean" of liquefied iron, nickel, and other elements. Currents, eddies, and rising bubbles of hot liquid metals sloshing around the outer core generate the Earth's protective magnetic field. The outer core is wrapped in a shell of liquefied rock called the mantle. The mantle is composed of much lighter elements than the metallic core, with silicon, aluminum, and oxygen dominating the hot, lava-like slush of the inner mantle. Further from the core, the mantle becomes rigid and semi-solid. The outer mantle, at a lower temperature, is a stiff, semi-hardened shell. Finally, the crust is the very thin, solidified surface of the planet. Over time, the crust will erode, break down, buckle, be resurfaced by lava, creating the changing, dynamic surface of the Earth.

Immediately upon emerging from the interior of the planet, we find the thin, gaseous shell of the Earth's atmosphere. Consisting mostly of nitrogen molecules (abundance: 78%) and oxygen molecules (abundance: 21%), the atmosphere of Earth also separates into distinct layers based on temperature and density. 99% of the gaseous particles in the Earth's atmosphere are located within 35 kilometers of the Earth's surface, in the layers known as the troposphere and stratosphere. Most weather systems arise in the troposphere, where blobs of warm air near the surface rise and cool air higher above the ground falls, giving rise to Earth's weather systems and meteorology. The stratosphere lies above the churning, mixing troposphere. The temperature of air in the stratosphere increases with altitude, leading to calm, generally turbulence-free conditions.

The mesosphere sees temperatures dropping with altitude again; it features the coldest temperatures in the atmosphere (about –230° F, or –140° C). The mesosphere is still not very well understood, as it lies far below the height of orbiting satellites and above the reach of weather balloons and aircraft. The density of the mesosphere is sufficient to vaporize many small incoming meteoroids, leading the streaks of light called "shooting stars" (meteors) across the sky. The northern lights – charged particles from the Sun mixing with and exciting gases in Earth's atmosphere – primarily originate in the mesosphere. The layer of gas lying above the mesosphere is called the thermosphere. High energy solar radiation – like UV light and X-ray photons – strongly excites the sparse atoms and molecules in the thermosphere, dumping energy into those and leading to temperatures of about 2500° C (4500° F) during the height of daylight hours (though to our naked skin, this would feel freezing cold because the few particles in the thermosphere would not be able to transfer sufficient amounts of heat). In the thermosphere, the very thin, sparse gases do not permit sound waves to propagate. Besides being the beginning of the silence of space, the thermosphere is also the beginning of "space" itself at approximately 100 kilometers. The International Space Station, Gemini and Mercury capsules, and the Space Shuttles orbited within the thermosphere.

Finally, the atmosphere peters out over the thousands and thousands of kilometers of the exosphere. Earth's gravity only weakly holds onto the very few, widely spaced particles in the exosphere, and in fact most of these constituents are washed away into interplanetary space by the Sun's solar wind. The relatively huge distance between atoms in the exosphere entails that these particles rarely collide with one another, and instead just loop and orbit around. The exosphere represents the very end of Earth's atmosphere.

PROCEDURE

On Worksheet #1, the Earth has been scaled down to one-ten-millionth of its actual size, so that every 1.0 centimeter on the page is equal to 1000 kilometers of the actual structure of the Earth. Each tick mark represents 200 km, with each increment of 1000 km labeled. The table titled "The Earth's Interior" includes the lower and upper boundaries of the Earth's main interior regions, from the solid iron inner core to the crust, which makes up the Earth's surface. Mark off the location of the upper boundary for each layer on the interior. Using a compass, draw a quarter semi-circle on the disk of the Earth to show the extent of each layer, placing the metal tip of the compass on the center of the Earth and using the pencil tip to trace out the curved upper boundary of the inner core, outer core, lower mantle, and upper mantle. Drawing the crust will not be possible due to its extremely small width at this scale.

Next, while referencing the second data table, mark off the position of the upper boundary of the stratosphere, mesosphere, thermosphere, and exosphere. The surface of the Earth is 0 kilometers (marked by a dotted line marking the end of the Earth's interior and the beginning of its atmospheric region). By using the compass, draw an arc around the Earth showing the extent of each layer.

Worksheet #2 focuses on the crust of Earth and the low atmosphere below the atmosphere-space boundary. If the full disk of the Earth was drawn, it would stretch over 20 feet across and require 2400 lab pages to completely display, so only a small curving sliver of the Earth's interior structure is shown. Each tick mark represents 2 km. Referring to the table titled "The Earth's Atmosphere," mark off the upper boundary altitude of the troposphere, the stratosphere, and the mesosphere. Using a ruler, draw a line across the page to show the extent of each of these layers.

Finally, using a small dot as a marker, label the locations of the highest sky dive, commercial jet liner cruising height, and height of cumulonimbus (large thunderstorm clouds). Also, using a ruler and a line, show the heights of the low ozone, high ozone, the Kármán line, and the region of the northern lights. The Kármán line is defined as the (somewhat arbitrary) boundary between the Earth's atmosphere and outer space. This definition is accepted by the Fédération Aéronautique Internationale (FAI), which is an international standard-setting and record-keeping body for aeronautics and astronautics. (Historically, in the U.S. the astronaut badge, also called astronaut wing, has been earned by a person flying higher than 80 km.)

Datasheet #1

The Earth's Interior		
Layer	**Lower Boundary (in km)**	**Upper Boundary (in km)**
Inner Core	0	1220
Outer Core	1220	3480
Lower Mantle	3480	5710
Upper Mantle	5710	6335
Crust	6335	6370

The Earth's Atmosphere		
Layer	**Lower Boundary Altitude (in km)**	**Upper Boundary Altitude (in km)**
Troposphere	0	12
Stratosphere	12	50
Mesosphere	50	80
Thermosphere	80	700
Exosphere	700	10000

Feature	**Altitude (in km)**
Cruising Height of Commercial Jets	12
Ozone Layer's Lower Boundary	20
Height of Cumulonimbus Clouds	25
Ozone Layer's Upper Boundary	30
Highest Skydive	40
Aurora Borealis (Northern Lights)	90
Kármán Line	100

Note: The various boundaries within Earth's atmosphere (denoted as "Altitude") within Earth's constitute approximate values.

Data obtained from:

https://www.nationalgeographic.org/encyclopedia/core/

https://www.nationalgeographic.org/encyclopedia/atmosphere/

Name _____ Id _____

Due Date _____ Lab Instructor _____ Section _____

Worksheet # 1

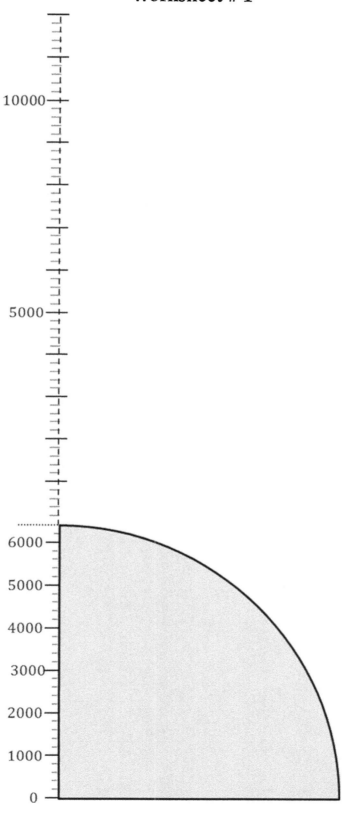

Name _____ Id _____

Due Date _____   Lab Instructor _____   Section _____

Worksheet # 2

UNIT 6.2   A SCALE MODEL OF THE SOLAR SYSTEM

OBJECTIVE
To demonstrate the various sizes and orientations of the Solar System's eight planets as well as lay out the groundwork for constructing a scale model Solar System extending out to Neptune

INTRODUCTION
The radius of a planet is a measure of a planet's size, from its uppermost solid layer (or the top of its atmosphere for a gas giant without a discernible solid surface) to the center of its core. Most of the planets are very close to spheres, but the rate at which a planet spins along with its composition causes the planets to bulge slightly at their equators. The most striking example is Saturn, with its very, very low density (about half the density of water) and very fast rotation rate (about 18,000 kilometers per hour) causing the planet's equator to bulge out 6000 kilometers more than its north pole to south pole distance.

Another pronounced characteristic of a planet is its axial tilt. The axial tilt of a planet describes the angle that the planet's equator makes to the planet's orbital path. For a planet with no axial tilt, the line of its north-south pole axis would be perpendicular (90°) to its orbital plane. The equator of an un-tilted planet would lie directly on top of the planet's orbital path. For various reasons – some lost to ancient history; several of the planets are substantially tipped and appear to lean toward or away from the Sun during their orbit, causing their north-south pole line to be bowed toward the path of their orbit. Earth's 23.5° tilt causes the rise of the seasons: when the Earth's orbital position leaves its north pole bowed toward the Sun, we experience the warmer summer months in the northern hemisphere; when Earth's pole is tilted away from the Sun, the northern hemisphere receives less light and experiences the cooler winter months.

Scale models are crucial to the world of science, engineering, architecture, and civil planning. A scale model is a perfectly proportioned miniature model of a much larger object. Before a building is constructed on campus, before the frame of a concept airplane is built, or before a car rolls off the assembly line, a physically smaller but proportionally identical model is created for testing and experimental purposes. The sizes and distances of Solar System objects make their scale difficult to grasp. Given the titanic size and scale of the Solar System, creating a properly scaled model is difficult. Either the planets are large and easily visible and the distances between the model planets are many miles across or the model is compact and small but the planets are microscopic.

However, in the same way as an architect may build a perfectly scaled miniature of a building to demonstrate what the structure will eventually look like, reproducing a scale model of the Solar System will give you insight into both the proportional size of the planets and their properly proportional distances. A scale model takes a size measurement and shrinks it down for all sizes and distances, equally. Unlike an elementary school science fair, where small Styrofoam balls all fit on a small table lined up with one another, you will quickly realize that even a scale model with a small scale can become incredibly large very quickly.

To calculate how large a scaled model must be, there is a need to relate real-world distances with scale model distances. In the case of a scale model, a planet with a diameter many thousands of kilometers wide may be shrunken down to just a few centimeters.

EQUATIONS AND CONSTANTS

Equation	Expression	Variables
Scale Factor	$Scale\ Factor = \dfrac{Real}{Ruler}$	*Scale Factor*: a conversion factor which bridges actual and scale model sizes *Real*: a real-world distance or size, such as kilometers or AU *Ruler*: a distance or size used in a scale model, such as meters or centimeters

ILLUSTRATIONS

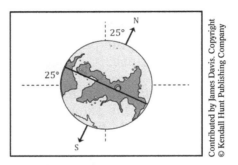

Figure 1. A schematic of the axial tilt of Mars. The dark black line represents Mars's equator with the N-S line representing the planet's polar axis, about which the planet rotates. The dashed horizontal line is the orbit of Mars, the route that it travels around the Sun, and it points directly at the Sun. The dotted vertical line is perpendicular to Mars's orbit and represents the location of Mars's north-south pole line if the planet had zero degrees of axial tilt. Instead, Mars is tilted 25° so its equator rises 25° above its orbital path and its axis likewise leans 25° away from the perpendicular line. The north-south spin axis and the equator are always perpendicular (i.e., they make a 90° angle to one another).

PROCEDURE

The first part of the lab involves drawing accurate representations of the eight planets, including their proper scaled diameters and axial tilts (i.e., their tilt measured with respect to the planetary orbital plane). Earth is 6,370 kilometers in radius, the distance from the core to the surface, with a diameter or 12,740 kilometers. In this model, 6,370 kilometers is set equal to 1.00 centimeter on paper, and all of the planets are drawn relative to this smaller scale. Your ScaleFactor, therefore, would be 6,370 km/cm. You may choose to use a different scale, however.

Using the Scale Factor equation and the actual radii of the eight planets and the Sun, calculate the scale model radius of the given objects, in centimeters. Draw these planets on Worksheets #3 through #7 using the steps listed below. On Worksheet #2, you will properly scale the distance between the miniature planets. Using the same scale factor as on Worksheet #1, convert the Sun-to-planet distances from kilometers into centimeters, using the same process as done on Worksheet #1. Since the numbers will be very large, convert the distances into meters. Using a map, computer program, or a list of locations and distances, choose a location for the center of the Solar System (the Sun). With the scale model distances, build a scale model of the Solar System by determining where the planets should be located (by naming a landmark or intersection that given distance from the Sun). To properly draw the size and orientation of the planets, follow the steps below and refer to the figures:

Step 1: On Worksheets #3 through #7, the orbital plane of each planet has been drawn and labeled. Place a very small **x** mark or dot on the dashed Orbit Line to mark the exact physical center of the planet. Using a ruler, draw a line from the **x** along the orbit with a length equal to the planet's scaled radius in centimeters from Worksheet #1.

Step 2: Using your compass, place the pencil end on the end of your radius line and the metal compass tip on the marked planetary center. Draw a smooth circle representing the planet's circumference. It is usually easier to actually hold the compass still and turn the paper, rather than trying to spin the compass itself around. The finished circle represents the full disk of each planet.

Step 3: Each of the planets has a distinct axial tilt. Using your protractor, place the center mark on the marked planetary center and place a small dot at the angle measurement corresponding to the planet's tilt. Draw a line across the planet's midsection. This is the geometrical equator of the planet and has a length equal to the scaled planet's diameter.

Step 4: Now line the protractor up on the equator – laid on the center mark again – and mark a small point at 90°. Draw a second straight line completely through the planet, perpendicular to the axis. This represents the north-south axis of the planet. Label the north pole with an N and the south pole with an S.

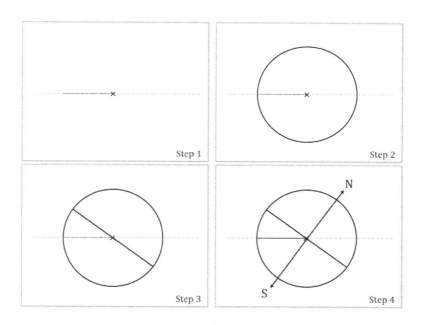

Name _____ Id _____

Due Date _____   Lab Instructor _____ Section _____

Worksheet # 1

Scale Factor Used (in km/cm)

Star	Radius (in km)	Scaled Radius (in cm)	Axial Tilt (in degrees)
Sun	695,500		0°

Planet	Radius (in km)	Scaled Radius (in cm)	Axial Tilt (in degrees)
Mercury	2,440		0°
Venus	6,050		177°
Earth	6,370		23.5°
Mars	3,390		25°
Jupiter	69,900		3°
Saturn	58,200		27°
Uranus	25,400		98°
Neptune	24,600		29°

Note: The axial tilt measures the angle between the plane of a planet's orbital plane and its equator.

Name _____ Id _____

Due Date _____   Lab Instructor _____   Section _____

Worksheet # 2

Planet	Distance (in km)	Distance (in AU)	Scaled Distance (in m)	Placement in Model
Mercury	57,910,000	0.387		
Venus	108,208,000	0.723		
Earth	149,598,000	1.00		
Mars	227,939,000	1.52		
Jupiter	778,547,000	5.20		
Saturn	1,433,449,000	9.58		
Uranus	2,870,671,000	19.2		
Neptune	4,498,542,000	30.1		

Name _____ Id _____

Due Date _____ Lab Instructor _____ Section _____

Worksheet # 3

The Inferior Planets: Mercury and Venus

Orbit
_____
Line

Orbit
_____
Line

Name _____ Id _____

Due Date _____   Lab Instructor _____   Section _____

Worksheet #4

The Outer Terrestrial Planets: Earth and Mars

Orbit
_____
Line

Orbit
_____
Line

Name _____ Id _____

Due Date _____   Lab Instructor _____   Section _____

Worksheet # 5

The Gas Giant: Jupiter

Orbit
_____

Line

Name _____ Id _____

Due Date _____   Lab Instructor _____   Section _____

Worksheet # 6

The Gas Giant: Saturn

Orbit
_____
Line

Name _____ Id _____

Due Date _____   Lab Instructor _____ Section _____

Worksheet # 7

The Ice Giants: Uranus and Neptune

Orbit
_____
Line

Orbit
_____
Line

Name _____ Id _____

Due Date _____   Lab Instructor _____   Section _____

Worksheet # 8

Postlab Questions

For each of the following questions, include all work, equations, and proper units.

1. The severity of seasonal temperature swings – summer to winter – is based on the tilt of a planet's axis toward (in summer) and away from (in winter) the Sun. The larger the tilt, i.e., the closer the poles lie to the orbital plane, the more extreme seasons are expected. Which *single planet* in the Solar System would have the most severe seasons in this case? Which *planets* would have the least severe seasons?

2. Saturn's brightest sets of rings, from one edge to the other, are 280,000 km wide. How many centimeters across would you draw the rings in your scale model?

3. The closest star to the Earth besides the Sun is the red dwarf star Proxima Centauri, located at a distance of 4.26 light years (269,000 AU). Using the same scale as used for the Solar System, how far away should Proxima Centauri be placed in a scale model (include units).

4. If 1 mile = 1610 meters, how far away should Proxima Centauri be from the scale model Sun (include units)?

5. The furthest apart one can place two objects on planet Earth is about 8,000 miles (two opposite sides of the planet). What is the immediate problem you see with building a scale model encompassing the Earth, planets, and Proxima Centauri?

UNIT 6.3   PLANETARY CHARACTERISTICS

OBJECTIVE
To learn about the physical parameters of terrestrial and Jovian planets such as diameter, mass, mean density and surface gravity

INTRODUCTION
With the demotion of Pluto in August 2006, the Solar System is considered to consist of the Sun, eight planets and a large number of smaller objects (natural satellites of planets, dwarf planets and other smaller objects, which include, e.g., asteroids, meteroids, and comets). The eight planets fall into two groups: the terrestrial planets (Mercury, Venus, Earth, Mars) and the Jovian (or gaseous) planets (Jupiter, Saturn, Uranus, Neptune). (Note however that the terms "Jovian planets" and "giant gas planets" are sometimes restricted to Jupiter and Saturn, and Uranus and Neptune are referred to as "medium-sized gas planets" or "ice giants".) These two planetary groups are characterized by significantly different physical properties with respect to their masses, volumes, densities, rotation rates, and composition. This unit will focus on relative comparisons of these properties within the two groups, as well as a comparison of properties between the groups.

The terrestrial planets: Earth is the largest and most dense terrestrial planet, with a mass greater than the mass of the other three terrestrial planets combined, although Venus's mass is about 82% of Earth's mass. Rather than using absolute units (like diameters in kilometers or masses in kilograms), it is intuitive to utilize "relative units" for describing the planets. For example, we may say the Moon's mass is 0.012 Earth Masses (1.2% = 1/81 of the mass of the Earth) rather than using its absolute mass value, which is 7.348×10^{22} kilograms. Earth's mass, gravitational pull, and diameter are used as the baseline against which the other planets are measured.

Calculate the relative diameter D_{rel}, relative mass M_{rel}, and relative surface gravity g_{rel} of the four major planets with Earth as the planet of reference. For data on the various planets, refer to Table 2 in the Appendix at the end of the manual. Give your results in the available table in Worksheet #1 and plot a histogram demonstrating the comparison between the distinct physical quantities of the terrestrial planets. The subscript "**E**" denotes the reference planet (Earth), and subscript "**pl**" denotes the respective planet, and the subscript "**rel**" denotes the relative ratio between the two.

$$D_{rel} = \frac{D_{pl}}{D_E} \qquad M_{rel} = \frac{M_{pl}}{M_E} \qquad g_{rel} = \frac{M_{pl}}{M_E} \times \left(\frac{D_E}{D_{pl}} \right)^2$$

The Jovian planets: Jupiter is the largest and most dominant planet in the Solar System. It is also typical of the diffuse, high mass, gaseous Jovian planets. All of the gas giants have ring systems (though none is as prominent as Saturn's ring system), fast rotation rates, high masses, and large diameters. Much as the terrestrial planets were compared to the Earth, compare the various gaseous planets to Jupiter, using the following relations for diameter, mass, and surface gravity:

$$D_{rel} = \frac{D_{pl}}{D_J} \qquad M_{rel} = \frac{M_{pl}}{M_J} \qquad g_{rel} = \frac{M_{pl}}{M_J} \times \left(\frac{D_J}{D_{pl}} \right)^2$$

PROCEDURE

Using the masses of the planets given in Table 2 of the Appendix, calculate the relative diameters, masses, and surface gravities of the eight planets (which have been separated into the terrestrial and Jovian planetary groups). Using the graphs, draw a histogram-style bar graph to make an accurate comparison of the planetary characteristics.

For each planet, mark off its diameter relative to the comparison planet and draw a line across the column and shade in the area beneath that line.

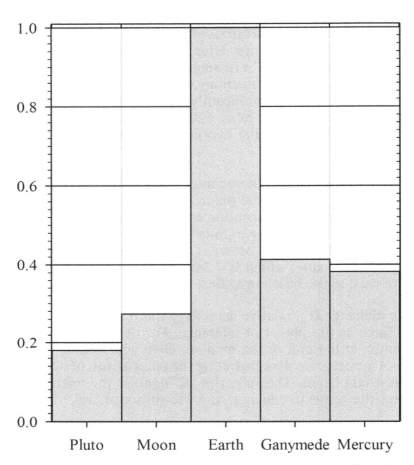

Relative Diameters

Figure 1. An example of a diameters histogram drawn for the Earth and four smaller Solar System bodies. The relative diameters of Pluto, the Moon, Ganymede (Jupiter's largest moon), and Mercury are depicted in comparison to the Earth. This comparative graph shows that Pluto is considerably smaller than the Moon and, furthermore, that Mercury – the smallest planet in the Solar System – is smaller than Ganymede, the largest moon in the Solar System.

Name _____ Id _____

Due Date _____   Lab Instructor _____ Section _____

Worksheet # 1

The Terrestrial Planets

Planet	Mass (in M_E)	Diameter (in D_E)	Surface Gravity (in g_E)	Density (in g/cm³)
Mercury				5.427
Venus				5.243
Earth	1.00	1.00	1.00	5.514
Mars				3.935

The Jovian Planets

Planet	Mass (in M_J)	Diameter (in D_J)	Surface Gravity (in g_J)	Density (in g/cm³)
Jupiter	1.00	1.00	1.00	1.326
Saturn				0.687
Uranus				1.270
Neptune				1.683

Name _____ Id _____

Due Date _____ Lab Instructor _____ Section _____

Worksheet # 2

Using the histogram style, graph the relative masses, diameters, surface gravities, and densities of the terrestrial planets:

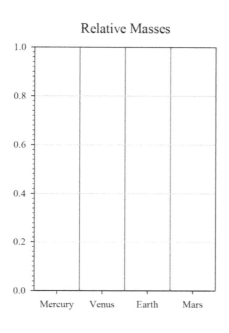

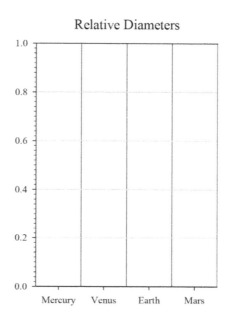

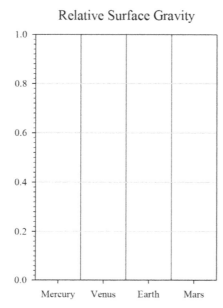

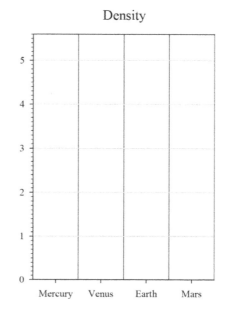

Name _____ Id _____

Due Date _____   Lab Instructor _____   Section _____

Worksheet # 3

Using the histogram style, graph the relative masses, diameters, surface gravities, and densities of the Jovian planets:

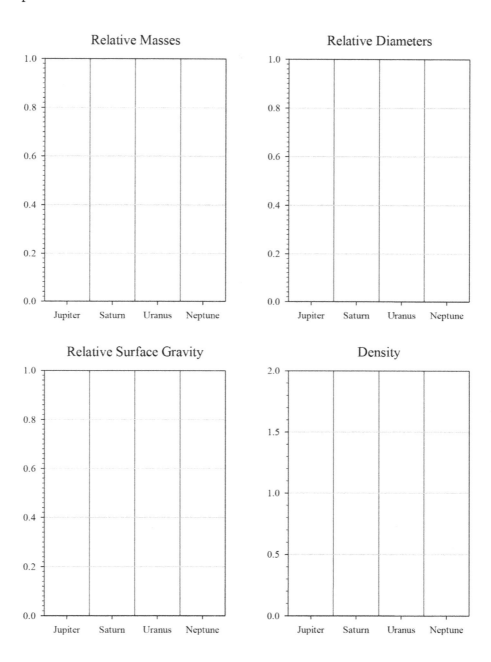

UNIT 6.4   MERCURY'S ORBIT

OBJECTIVE

To learn how observations of Mercury, including measurements of its angular separation from the Sun, allowed astronomers to determine the extent of Mercury's orbit, which helped validating the heliocentric model of the Solar System

INTRODUCTION

Planets appear as little more than bright points of light in the night sky. They attracted the attention of ancient sky watchers because of their "wandering" nature, in contrast to the unchanging constellations. The Greek word for planet directly translates as "wandering star," as the planets appear to change position from one night to the next, moving from one constellation to another over periods of months or years. For this reason, every sky-watching culture recognized that these objects were fundamentally different from the stars. Before the advent of the telescope, astronomers used instruments such as the quadrant and the sextant to measure planetary positions. These instruments had no optics, no ability to magnify, no ability to brighten an object or to produce photographs. They were essentially surveying equipment, allowing the user to plot very accurate, very precise star charts of celestial positions as they appeared in the night sky.

The seven "planets" recognized by ancient astronomers were Mercury, Venus, Mars, Jupiter, Saturn, the Sun (which was not recognized as a star) and the Moon (which was not recognized as the Earth's natural satellite). Furthermore, the planets Mercury and Venus were referred to as "inferior" planets. The idea of "inferior" denoted that these two planets – for reasons unclear to astronomers who believed in an Earth-centered Universe – were in an orbit that always kept them in the same general region of the sky as the Sun. Mercury could never be seen more than 28° from the Sun, while Venus – at most – could only be 47° degrees from the Sun. The planets Mars, Jupiter, and Saturn, on the other hand, could appear in any part of the sky, regardless of the position of the Sun.

When the inferior planets appear farthest from the Sun, this is called "greatest elongation." At greatest elongation, the measured angle between the Sun and the planet is at the largest possible extent. A "greatest eastern elongation" means that the planet is as far to the east of the Sun as possible and thus will appear in the evening sky, setting after the Sun. A "greatest western elongation" occurs when the planet is far to the west of the Sun, appearing in the morning sky and rising before the Sun. Mercury and Venus at western elongation were referred to as morning stars, as they were sometimes still visible even after the light of the rising Sun had obscured the night's constellations.

A planet's orbit may be characterized by several features: its eccentricity, aphelion distance, perihelion distance, major axis, and semi-major axis. The eccentricity describes the flatness of an orbit. An orbit with an eccentricity of 0.0 is a perfect circle, with equal distances from the center to every point on the orbit. For a planet orbiting the Sun, "perihelion" is the point in orbit when the planet is closest to the Sun, with 'peri' the Greek stem for 'near' and 'helion' the word for Sun. The opposite, "aphelion," is the furthest point in orbit from the Sun. The perihelion and aphelion point are always directly opposing one another.

The major-axis is the length of the orbit along its longest axis, from the perihelion point to the aphelion point. The major axis is thus the distance from the perihelion point to the Sun to the aphelion point (which should all align with one another). For a perfect circle, the major axis would be called the diameter. The semi-major axis, which may be considered the 'radius' of an ellipse, is half the length of the major-axis (making it the distance from the perihelion point or the aphelion point to the dead center of the orbit, which is not the position of the Sun, in most cases).

EQUATIONS AND CONSTANTS

Equation	Expression	Variables
Eccentricity	$e = \dfrac{Q-q}{Q+q}$	Q: the aphelion distance q: the perihelion distance
Semi-Major Axis	$a = \dfrac{Q+q}{2}$	Q: the aphelion distance q: the perihelion distance
Distance in AU	$D = \dfrac{a_{planet}}{a_{Earth}}$	a_{planet}: semi-major axis of the planet in any distance unit a_{Earth}: semi-major axis of Earth as measured in the same distance unit as a_{planet}

ILLUSTRATIONS

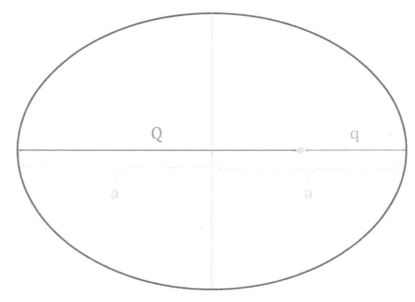

Figure 1. A schematic drawing of an elliptical orbit with the perihelion distance (smallest planet-Sun distance) labeled as q, the largest planet-Sun distance (aphelion) labeled as Q, and the semi-major axis (the average separation between the Sun and the planet) labeled as a. The major axis is the distance from the perihelion point to the aphelion point.

PROCEDURE

The circle on Worksheet #1 represents the Earth's orbit about the Sun (the central dot), with each of the long tick marks (with the circles at the end to identify them) representing the first day of that particular month, the short, dashed lines representing every 5th day of the month, and the solid, medium-length tick mark representing the 15th day of the month (about the mid-point of the month). For each of the dates given in Datasheet #1, find the point on Earth's orbit closest to that date and mark that position with the data point number given in the table. You may choose to draw a very light line from the Sun to that date on the Earth's orbit to serve as a guide. Orient Worksheet #1 so that the position you marked is directly in front of you.

Next, determine the position and direction of planet Mercury, given in Datasheet #1. Place the center mark of your protractor on the Earth's position and mark off the appropriate number of degrees either to the left of the Sun (for eastern elongation) or right of the Sun (for western elongation) as indicated in the table. It is critical to the rest of the lab that you have the protractor oriented correctly. Draw a long, dark line along this line of sight to the mark. This line is a tangent to Mercury's orbit, touching it in only one place. Taken with all of the other readings, a pattern of tangent lines will emerge that will trace out Mercury's orbit around the Sun.

Sketch the orbit as best as possible as an ellipse that touches all of the tangent lines without crossing them or falling short. There will be rough edges to the graph, but if you had drawn the map with hundreds or thousands of lines, the smooth, continuous orbit would be revealed. An estimated sketched orbit based on approximately 20 points should work well and give you all the data needed for calculating Mercury's orbit. Using a ruler, measure the orbit and find the perihelion point (closest point from Mercury's orbit to the center of the Sun) and mark it with a **q**. Find the aphelion point in the same way (hint: it should fall directly across from your perihelion point) and mark it with a **Q**. Use a ruler and measure to ensure that those are the longest / shortest separations.

Datasheet #1

The following table lists the greatest elongations of the planet Mercury over several years. The western elongations make an angle with Mercury to the right of the Sun as seen from the Earth. The eastern elongations make an angle to the left of the Sun as seen from the Earth

Data Point #	Year	Month	Date	Position Angle (°)		Direction
1	2014	January	31	18.4		East
2	2014	March	14		27.6	West
3	2014	May	25	22.7		East
4	2014	July	12		20.9	West
5	2014	September	21	26.4		East
6	2014	November	1		18.7	West
7	2015	January	14	18.9		East
8	2015	February	24		26.7	West
9	2015	May	7	21.2		East
10	2015	June	24		22.5	West
11	2015	September	4	27.1		East
12	2015	October	16		18.1	West
13	2015	December	29	19.7		East
14	2016	February	7		25.5	West
15	2016	April	18	19.9		East
16	2016	June	5		24.2	West
17	2016	August	16	27.4		East
18	2016	September	28		17.9	West
19	2016	December	11	20.8		East
20	2017	January	19		24.1	West
21	2017	April	1	19.0		East

Name _____ Id _____

Due Date _____   Lab Instructor _____   Section _____

Worksheet # 1

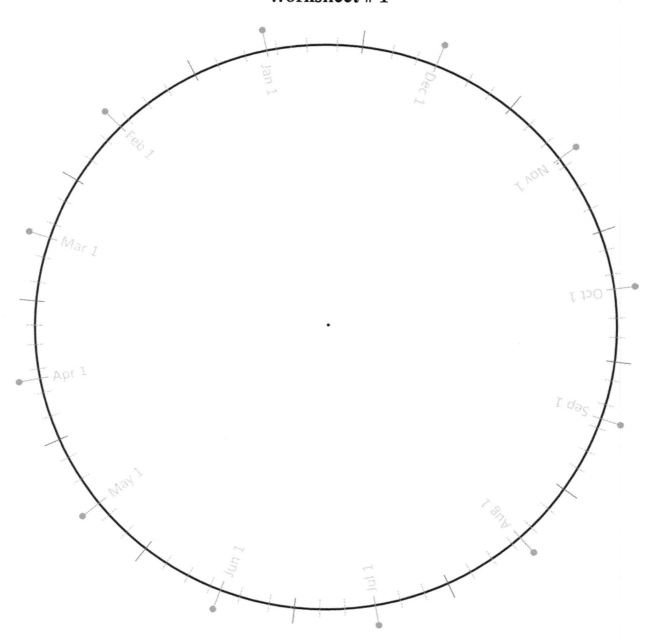

Name _____ Id _____

Due Date _____   Lab Instructor _____   Section _____

Worksheet # 2

For each of the following questions, where a numerical answer is required, clearly show what equation you are using, what numbers you are plugging into the variables, and what the units are.

1. The average distance from the Earth to the Sun (the semi-major axis of Earth's orbit) is 1.0 AU. In the scaled-down map of the Earth's orbit on Worksheet #1, use a ruler and measure the distance from the Earth's orbit to the Sun. How many millimeters represent 1.0 AU on that diagram? (Measure anywhere on the orbit: the eccentricity of Earth's orbit is immeasurably small in this diagram.)

2. On Worksheet #1, identify the perihelion point of Mercury's orbit and mark it with a **q**. What is the distance from Mercury to the Sun at Mercury's perihelion in millimeters on your diagram? How many real-world AU does this translate into?

3. On Worksheet #1, identify the aphelion point of Mercury's orbit and mark it with a **Q**. What is the distance from Mercury to the Sun at Mercury's aphelion in millimeters on your diagram? How many real-world AU does this translate into?

4. By using your above measurements, calculate the semi-major axis of Mercury's orbit in AU. Show your work.

5. What is the eccentricity of Mercury's orbit? Show your work.

6. The former planet Pluto draws as close to the Sun as 29.1 AU at perihelion and as far from the Sun as 49.2 AU at aphelion. What is Pluto's eccentricity?

7. If you were to draw Pluto's orbit using the same scale as you did for Mercury, how far from the Sun would you have to draw Pluto at its aphelion point (given in mm)?

UNIT 6.5   KUIPER BELT OBJECTS

OBJECTIVE

To learn about the properties of the Kuiper Belt Objects orbiting beyond Neptune, including their physical sizes, orbital eccentricities, and orbital inclinations

INTRODUCTION

Following the discovery of Uranus in 1781 and Neptune in 1846, astronomers searched for signs of other planets lurking in far reaches of the Solar System on long-period orbits for centuries. It has not been until Clyde Tombaugh, working at the Lowell Observatory (Flagstaff, Arizona), spotted a small pinpoint of moving light on a photographic plate in 1930. A new planet exterior to the orbit of Neptune had been discovered. The astronomer Gerard Kuiper theorized that the formation of the Solar System would leave a ring of icy, leftover material in the far reaches of the Solar System, from just beyond Neptune's orbit (~30 AU) out to about 55 AU. Like the asteroid belt between Mars and Jupiter, the Kuiper Belt would be a wide disk of planetary debris slowly orbiting the Sun. It was from this location (aside from the Oort Cloud) that the Solar System's comets would originate: small chunks of ice and rock that vaporize spectacularly as they approach the Sun. Kuiper theorized that many billions of tiny icy fragments may orbit out in that vast region, with a comet created every time a collision or gravitational nudge disturbed the orbit and caused it to careen toward the Sun.

The first Kuiper Belt Object (KBO) fitting this description was 1992 QB_1, discovered in 1992, showing for the first time that there were objects in the outskirt of the Solar System, inhabiting the same region of space as Pluto. In 2005, the object 2003 UB_{313} was identified to be about the same size as but more massive than Pluto, eventually dubbed Eris. Eris joined other substantial KBOs, like Makemake and Haumea. Taken together, these objects showed that Pluto composed only a tiny fraction of the Kuiper Belt's mass, representing a less dominant object than any of the Solar System's eight planets and more a primitive, unfinished remnant of planetary formation. Pluto was subsequently reclassified from planet to dwarf planet.

Today, over 1000 KBOs have been confirmed, with enough observations to allow astronomers to plot out and predict their orbits. The simple Kuiper Belt turned out to be more complex than originally thought, consisting of different, distinct families of KBOs. The objects called Plutinos (like, nor surprisingly, Pluto) show effects of being strongly influenced by Neptune's gravity. Plutinos have perihelion distances which are very close to Neptune's semi-major axis distance of 30.1 AU. In addition, Plutino orbits synchronize with Neptune's orbit, with lengths that are 1.50, 2.00, or 2.50 times the length of Neptune's 164.8 year orbit (i.e., orbits of 248 years, 330 years, or 412 years, respectively).

A second class of KBO, called Cubewanos (like the namesake object QB_1) have orbits that are unperturbed and unaffected by Neptune. Their eccentricities are very small, usually less than 0.1, and have perihelion distances that are nowhere near Neptune's 30.1 AU semi-major axis. Finally, the Kuiper Belt is populated by some objects showing a history of having been badly disturbed into elongated, chaotic orbits by interactions with Neptune or some other distant object. These bodies are called Scattered Disk Objects (SDOs). Scattered disk KBOs are marked by high inclinations, with tilted orbits that lie high above or below the orbits of the eight planets, along with aphelion and perihelion distances that vary greatly and never get very close to Neptune's orbit (with eccentricities larger than 0.25). The scattered disk serves as evidence that the early Solar System was a more disturbed, chaotic place, with these KBOs serving as reminders of the past upheaval.

EQUATIONS AND CONSTANTS

Equation	Expression	Variables
Aphelion	$Q = a(1+e)$	Q: the aphelion distance a: the semi-major axis of the orbit e: the eccentricity of the orbit
Perihelion	$q = a(1-e)$	q: the perihelion distance a: the semi-major axis of the orbit e: the eccentricity of the orbit
Kepler's Third Law	$a^3 = P^2$	a: the semi-major axis of the orbit in AU P: the period of the orbit in years

PROCEDURE

For Worksheet #1, you will identify the largest KBOs by their size. Refer to Datasheet #1 as well as Table 3 in the Appendix for information. The largest circle (marked by a dotted line) represents the circumference of Mercury, the Solar System's smallest planet. The other circles (labeled 1 through 11) represent the 11 largest known KBOs (including Pluto's largest moon, Charon). Despite its small size, Mercury's diameter of 4880 kilometers is over twice the diameter of the largest KBO, Pluto. The drawings of the KBOs' circumferences are scaled so that 1.0 millimeter in the drawing is equal to 36 kilometers of actual length. Because many KBOs have remarkably similar diameters, which appear almost identical in the drawings, several of the objects have already been identified on Worksheet #1. By measurement and elimination, determine which circles correspond to the physical sizes of the remaining 7 KBOs.

Also on Worksheet #1, choose 5 KBOs (excluding Sedna and Pluto's moon Charon). Enter their names. From the Appendix, enter the semi-major axis of the KBO and its eccentricity and calculate both its aphelion distance (in AU) and its perihelion distance (in AU).

Part of Pluto's reclassification from planet to dwarf planet was based on its inability to "clear its neighborhood" of other KBOs. This means that Pluto's orbital region overlaps with the orbital regions of numerous other KBOs. On Worksheet #2 there are five columns, each with a space below to record the name of a KBO from Worksheet #1 and with an axis labeled with distances running from 0 AU (the location of the Sun) to 100 AU (far beyond the outer edge of the Kuiper Belt). The shaded horizontal boxes represent the aphelion and perihelion distances of the eight Solar System's planets, showing the domain dominated by them. Note that Uranus is the most eccentric planet, with a widely spaced perihelion and aphelion distance. Also note that none of the eight planets' domains overlap (or even draw close to one another). Using your KBOs and their calculated aphelion and perihelion distances from Worksheet #1, plot a pair of lines across the single column to represent the perihelion and aphelion distance of each KBO. Shade in the space between those lines. This represents the "neighborhood" of the KBO. For a planet, that neighborhood would be free of any overlapping neighborhoods from the other columns. This will clearly not be the case for the KBOs, as you may notice significant regions of overlap.

For Worksheet #3, use your five chosen KBOs again and look up their orbital inclinations (the tilt of their orbit relative to the plane of the Earth's orbit around the Sun). Mark with a small dot the angular position of those KBO's orbits, connect the dot to the location of the Sun with a dashed line (representing the plane of the planet's orbit) and label the point with the KBO's name.

Datasheet #1

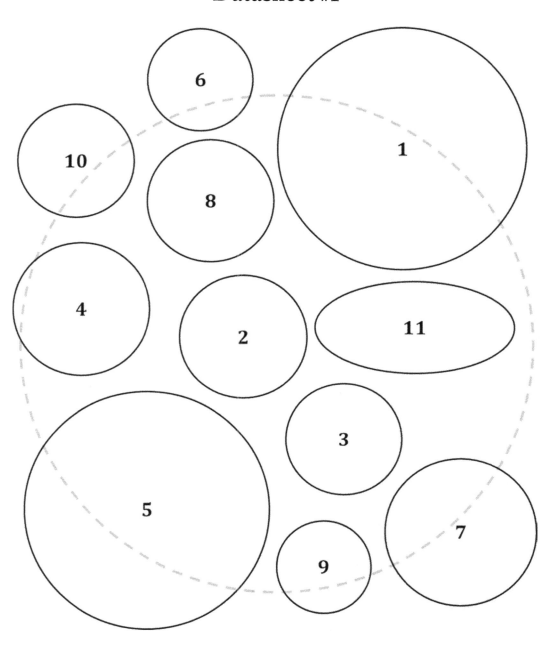

Name _____ Id _____

Due Date _____   Lab Instructor _____   Section _____

Worksheet # 1

KBO Sizes

Object	KBO Name	Measured Diameter (in mm)	Calculated Diameter (in km)
1	Pluto	66	2370
2	Charon	34	1200
3	Orcus	31	1110
4	2007 OR$_{10}$	36	1290
5			
6			
7			
8			
9			
10			
11			

KBO Orbits

KBO Name	Semi-Major Axis (in AU)	Eccentricity	Perihelion (in AU)	Aphelion (in AU)

Data obtained from:

http://www2.ess.ucla.edu/~jewitt/kb/big_kbo.html

Name _____ Id _____

Due Date _____ Lab Instructor _____ Section _____

Worksheet # 2

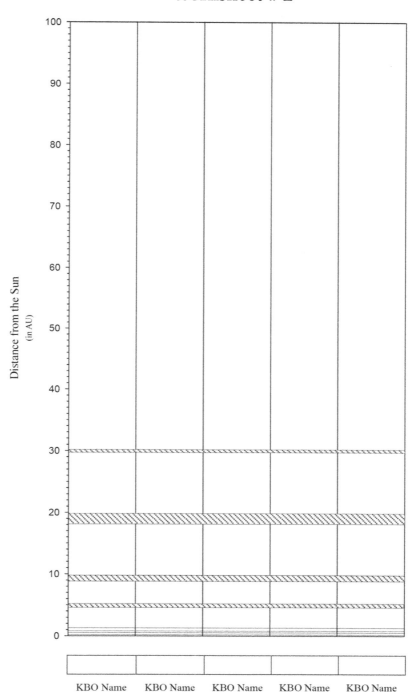

Name _____ Id _____

Due Date _____   Lab Instructor _____   Section _____

Worksheet # 3

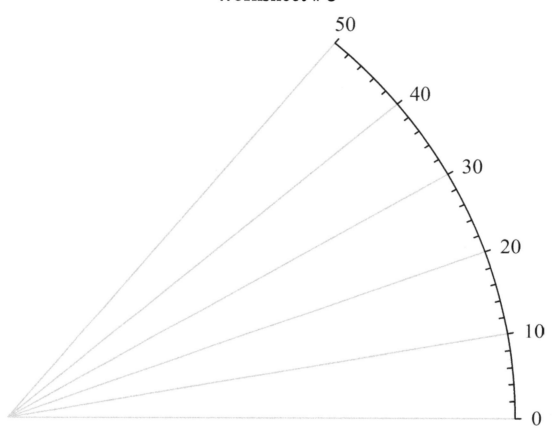

Name _____ Id _____

Due Date _____   Lab Instructor _____ Section _____

Worksheet # 4

KBOs fall into three classes, as covered in the Introduction: cubewanos, plutinos, and SDOs. Using the data for the KBOs from the Appendix Table 3, answer the following questions about the nature of some KBOs. Calculate the necessary values of perihelion, period, and period-ratio or state pertinent values given in the KBO data table, such as inclination or eccentricity. Points are awarded on thoroughness of answers, so be as complete as possible in your replies.

1. Referring back to the orbital inclinations of your KBOs as selected, why even if two KBOs have overlapping "neighborhoods" or share an orbit, they will be very unlikely to crash into one another?

2. What family of KBOs does Orcus belong to? Is it a Cubewano, a Plutino, or an SDO? Show calculations or measurements that lead to this conclusion and explain your answer.

3. What family of KBOs does Quaoar belong to? Is it a Cubewano, a Plutino, or an SDO? Show calculations or measurements that lead to this conclusion and explain your answer.

4. What is Eris's orbital period, in years?

5. Today, Sedna is about as close to the Sun as its orbit allows it to be, which allowed astronomers to detect it before its long orbit takes it far out of the Kuiper Belt and makes it invisible to most telescopes. What is Sedna's period, in years?

6. In 2014, astronomers discovered 22 new KBOs, among those the following three: 2014 UF_{224}, with a period of 331.1 years; 2014 TT_{85}, with a period of 280.8 years; 2014 QM_{441}, with a period of 315.2 years. Given that information and knowing that Neptune's period is 164.8 years, which of those objects – if any – are likely part of the Plutino family of KBOs? Show your work that lead you to your conclusion.

UNIT 6.6   EXOPLANETARY SYSTEMS AND HABITABLE ZONES

OBJECTIVE

To learn about the nature of planetary systems outside of the Solar System, including learning about what is required (based on a tentative and preliminary understanding) for an Earth-like planet to be considered suitable for life

INTRODUCTION

Our Solar System is composed of a single star, eight planets (four heavyweight gas giants and four compact, dense terrestrial planets), as well as rocky and icy debris in the form of asteroids, comets, and Kuiper Belt Objects. While it had long been theorized that other stars could host their own planetary systems, evidence proved elusive until the discovery of a large, Jupiter-sized planet orbiting the star 51 Pegasi in 1995. Thereafter, astronomers have utilized various planet-hunting techniques to confirm the existence of thousands of planets around other stars, commonly referred to as extrasolar planets or exoplanets.

Using statistical analysis of the planets already discovered – their numbers, ages, and types of stars they orbit – astronomers estimate that the Milky Way Galaxy alone may be home to over 100 billion planets, from the gigantic (ten times heavier than Jupiter) to the tiny (smaller than Mercury); from the incredibly hot (3000 K, hot enough to liquefy iron) to the cold (10 K, nearly absolute zero). These planets have been found orbiting all types of stars: large and small, hot and cool, young, highly evolved, and even extinct. Planets have even been found "orphaned," ejected from their orbit around a star and floating through the Galaxy alone.

Of particular interest to astronomers today is the search for a twin to Earth: a planet with a similar mass, radius, temperature, and atmospheric composition. With modern telescopic observations and computer models, planetary masses and diameters can be calculated with accuracy and precision. Instruments are not to the point yet where the chemical makeup of a planet's atmosphere can be easily determined, but the distance from star to planet (semi-major axis of the planet's orbit), the mass of the planet (the amount of matter making up the bulk of the planet's body), and the physical diameter of the planet can be measured with confidence. Since the first few hundred planets discovered were gas giants, it became a standard procedure to measure the mass of planets in terms of Jupiter masses (with 1 Jupiter mass = 318 Earth masses).

Any planet with a mass greater than 0.25 M_J (Jupiter masses) is usually considered a gas giant – a huge, low density planet with a large radius dominated by a tens-of-thousands of miles thick atmospheres of hydrogen and helium. While these planets could have habitable moons, the planets themselves are radically different from Earth and not suitable for life. Any planet with a mass smaller than 0.25 M_J and larger than 0.03 M_J is considered Neptune-like: still a huge gas giant, still dominated by a thick hydrogen-helium-methane-ammonia atmosphere with (perhaps) a solid rocky surface beneath the atmosphere as well as possibly a scalding hot ocean of water and slush (the same as Neptune and Uranus in our Solar System). These objects are almost certainly uninhabitable as well.

Finally, any planet with a mass less than 0.015 M_J can be considered "Earth-like" (in a broad sense of this term) as it constitutes a small, rocky-terrestrial type body. While such a planet may still be utterly unsuitable for life – lacking an atmosphere, like Mercury, or with a crushing sulfuric acid and carbon dioxide atmosphere like Venus – it at least is physically type-of-similar to Earth in size and probably internal structure.

Not only must a planet be roughly the size of Earth to be considered habitable, it must also exist in a position where the amount of starlight it receives is adequate to keep the planet temperate. Specifically, since we don't know of any life forms that can exist without the presence of liquid water, astronomers assume that to be habitable, a planet must have liquid water on its surface. Too close to the star and the incoming sunlight will vaporize surface water and leave the planet a scalding hot greenhouse. Too far from the star and any surface water would freeze completely and permanently. The area of space around a star where an Earth-like planet could have any hope of maintaining liquid surface water is called the habitable zone (or sometimes the Goldilocks zone – not too hot, not too cold, and just right for liquid water to exist on the planetary surface).

The habitable zone is strongly dependent on the luminosity of a star: the combination of the star's size and temperature. A highly luminous star will have a habitable zone much further away than the Sun's habitable zone. A low luminosity star will have a habitable zone located much closer to the star's surface than in our Solar System. The luminosity of a star is impossible to measure directly but a star's size and temperature may be measured, allowing astronomers to confidently calculate the luminosity. With that in hand, astronomers can then set boundaries on the habitable zone and determine if any of the host planets reside within.

EQUATIONS AND CONSTANTS

Equation	Expression	Variables
Luminosity	$L = R^2 \left(\dfrac{T}{5780} \right)^4$	L: a star's luminosity in solar luminosities R: a star's radius in solar radii T: a star's temperature in Kelvin
Inner Habitable Zone Distance	$D_{in} = \sqrt{\dfrac{L}{1.1}}$	D_{in}: the distance from the star to the habitable zone's inner boundary in AU L: the star's luminosity in solar luminosities
Outer Habitable Zone Distance	$D_{out} = \sqrt{\dfrac{L}{0.53}}$	D_{out}: the distance from the star to the habitable zone's outer boundary in AU L: the star's luminosity in solar luminosities

PROCEDURE

Datasheet #1 contains a list of 9 exosolar planetary systems. The mass, temperature, and radius of the central star is listed, along with the best estimates for each individual planet's mass (in Jupiter masses) and its distance from the central star (the semi-major axis of its orbit in AU).

Choose any 4 systems and use the given data and the equations to calculate the star's luminosity (in units of solar luminosity, $L_\odot$), the distance from the star where the habitable zone begins (D_{in}, measured in AU), and the distance from the star where the habitable zone ends (D_{out}, measured in AU).

On Worksheet #1, record the name of the star system (i.e., 55 Cancri, Kepler 62, etc.), the central star's radius, temperature, luminosity, and the location of the habitable zone's inner and outer boundaries. In the spaces provided, also record the distance between the star and its planets.

On Worksheet #2, plot the position of the planets and the boundaries given by the inner and outer habitable zone. The central star is located at 0.0 AU. The tick-marks on the graph are increments of 0.1 AU. Use a pair of vertical lines to mark the locations of the beginning and end of the habitable zone, labeling each. Any planet located between the inner and outer zones may be habitable worlds.

Carefully plot the location of the star's planets, as well, marking each planet with a small dot and labeling it with its designating letter (*b, c, d, e,* or *f*). In the example at the top of Worksheet #2, our Solar System is plotted, with Mercury (M), Venus (V), Earth (E), and Mars (M) marked at their respective distances from the Sun (0.39 AU, 0.72 AU, 1.00 AU, and 1.52 AU, respectively). Considering the location of the Sun's inner and outer habitable zone boundaries, only one Solar System's planet falls within the Goldilocks zone (Earth, not surprisingly).

Finally, use your graphs and data tables as well as information from the Introduction to answer the postlab questions.

Datasheet #1

The data tables below include a small selection of the Milky Way's known exoplanets.

Planet Name	Stellar Mass (in $M_\odot$)	Stellar Temperature (in K)	Stellar Radius (in $R_\odot$)	Planet's Distance from Star (in AU)	Planet's Mass (in M_J)
55 Cnc b	0.905	5196	0.943	0.1134	0.8
55 Cnc c	0.905	5196	0.943	0.2403	0.169
55 Cnc d	0.905	5196	0.943	5.74	3.82
55 Cnc e	0.905	5196	0.943	0.0156	0.02618
55 Cnc f	0.905	5196	0.943	0.781	0.144
Kepler 62 b	0.69	4875	0.62	0.0553	0.028
Kepler 62 c	0.69	4875	0.62	0.0929	0.0126
Kepler 62 d	0.69	4875	0.62	0.120	0.0440
Kepler 62 e	0.69	4875	0.62	0.427	0.0112
Kepler 62 f	0.69	4875	0.62	0.718	0.110
61 Vir b	0.95	5530	0.94	0.0502	0.016
61 Vir c	0.95	5530	0.94	0.2175	0.0573
61 Vir d	0.95	5530	0.94	0.476	0.072
HD 190360 b	1.04	5588	1.2	3.92	1.502
HD 190360 c	1.04	5588	1.2	0.128	0.057
Gliese 576 b	0.334	3350	0.36	0.28317	1.927
Gliese 576 c	0.334	3350	0.36	0.12959	0.637
Gliese 576 d	0.334	3350	0.36	0.02081	0.017
Gliese 576 e	0.334	3350	0.36	0.3343	0.039
HD 37124 b	0.83	5610	0.82	0.53364	0.675
HD 37124 c	0.83	5610	0.82	1.71	0.652
HD 37124 d	0.83	5610	0.82	2.807	0.696
HIP 14810 b	0.99	5485	1.00	3.88	0.0692
HIP 14810 c	0.99	5485	1.00	1.28	0.545
HIP 14810 d	0.99	5485	1.00	0.57	1.89
Gliese 667C b	0.31	3700	0.42	0.0504	0.01718
Gliese 667C c	0.31	3700	0.42	0.1251	0.0134
Gliese 667C d	0.31	3700	0.42	0.3035	0.0218

Continue....

Planet Name	Stellar Mass (in M$_\odot$)	Stellar Temperature (in K)	Stellar Radius (in R$_\odot$)	Planet's Distance from Star (in AU)	Planet's Mass (in M$_J$)
Gliese 667C e	0.31	3700	0.42	0.213	0.0085
Kepler 102 b	0.81	4900	0.74	0.0554	0.0135
Kepler 102 c	0.81	4900	0.74	0.0672	0.00944
Kepler 102 d	0.81	4900	0.74	0.0864	0.012585
Kepler 102 e	0.81	4900	0.74	0.117	0.028
Kepler 102 f	0.81	4900	0.74	0.166	0.0164

Data obtained from:

http://exoplanet.eu/

https://exoplanets.nasa.gov/

Name _____ Id _____

Due Date _____ Lab Instructor _____ Section _____

Worksheet # 1

Choose 4 planetary systems from the data sheet and fill in the information below. Use the equations from the lab to calculate the star's luminosity, the distance to the inner habitable zone and the distance to the outer habitable zone. The other information, including star name, radius, and planetary distances, are located in the data table.

Star System # 1:

Central Star Name	

Stellar Radius (in $R_\odot$)	Stellar Temperature (in K)	Stellar Luminosity (in $L_\odot$)	Inner Habitable Zone Distance (in AU)	Outer Habitable Zone Distance (in AU)

Distance to Planet #1 (in AU)	Distance to Planet #2 (in AU)	Distance to Planet #3 (in AU)	Distance to Planet #4 (in AU)	Distance to Planet #5 (in AU)

Star System # 2:

Central Star Name	

Stellar Radius (in $R_\odot$)	Stellar Temperature (in K)	Stellar Luminosity (in $L_\odot$)	Inner Habitable Zone Distance (in AU)	Outer Habitable Zone Distance (in AU)

Distance to Planet #1 (in AU)	Distance to Planet #2 (in AU)	Distance to Planet #3 (in AU)	Distance to Planet #4 (in AU)	Distance to Planet #5 (in AU)

Name _____ Id _____

Due Date _____ Lab Instructor _____ Section _____

Worksheet # 2

Star System # 3:

Central Star Name	

Stellar Radius (in $R_\odot$)	Stellar Temperature (in K)	Stellar Luminosity (in $L_\odot$)	Inner Habitable Zone Distance (in AU)	Outer Habitable Zone Distance (in AU)

Distance to Planet #1 (in AU)	Distance to Planet #2 (in AU)	Distance to Planet #3 (in AU)	Distance to Planet #4 (in AU)	Distance to Planet #5 (in AU)

Star System # 4:

Central Star Name	

Stellar Radius (in $R_\odot$)	Stellar Temperature (in K)	Stellar Luminosity (in $L_\odot$)	Inner Habitable Zone Distance (in AU)	Outer Habitable Zone Distance (in AU)

Distance to Planet #1 (in AU)	Distance to Planet #2 (in AU)	Distance to Planet #3 (in AU)	Distance to Planet #4 (in AU)	Distance to Planet #5 (in AU)

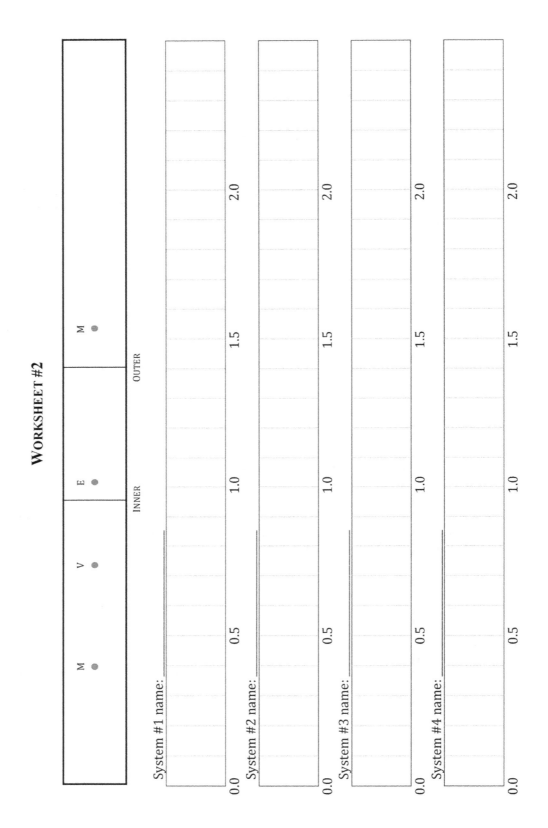

Name _____ Id _____

Due Date _____   Lab Instructor _____   Section _____

Worksheet # 3

For each of the following questions, where a numerical answer is required, clearly show what equation you are using, what numbers you are plugging in as variables, and what are the units of your final answer.

1. Planet masses are measured in terms of Jupiter masses, the Solar System's largest planet. Jupiter is located a distant, cold, 5 AU from the Sun. How many "hot Jupiters" (if any) does your exosolar systems contain? Hot Jupiters shall be defined as planets with masses greater than 0.25 Jupiter masses and distances smaller than D_{in}, the inner limit of the stellar habitable zone.

2. While astronomers cannot yet reliably discern (at least in most cases) the chemical make-up or density of exoplanetary atmospheres, they can measure distances from the central star and sometimes they can determine exoplanetary masses. Given that Earth's mass is 0.003 Jupiter masses and Earth exists within the central star's habitable zone, how many of your planets are possibly habitable Earth-like planets? (Earth-like masses extend up to 0.015 M_J)

3. Gas giants – like Jupiter, Saturn, Uranus, and Neptune – often have multiple moon systems (from Jupiter's 67 moons to Neptune's 13). Some of these moons are large, spherical, and icy. If their temperatures were high enough, the icy crusts would melt into global oceans, leaving the planets with one of the most critical components of life: liquid water. How many planets in the four chosen systems could host ocean moons (gas giants in the habitable zone).

4. In total, in your four exosolar systems, how many possibly habitable planets might exist? (Count each gas giant system as one habitable world, though certainly a gas giant could have multiple habitable moons.)

UNIT 7: FROM STARS TO GALAXIES

UNIT 7.1   PARALLAX: THE DISTANCE TO A STAR

OBJECTIVE

To learn how astronomers use observations and geometry to calculate the distance to stars. You will learn how to measure a stellar parallax and turn this measurement into a physical distance in space. This is the first rung on the distance ladder which allows astronomers to determine the distances to increasingly far-flung objects in the Universe like far-away star clusters and galaxies.

INTRODUCTION

"Depth perception" is the visual ability to interpret flat, two-dimensional images as being three-dimensional. Depth perception allows for the perception of distance. Each eye receives light from the environment, that light falls onto specific light sensors in the eye, and those stimulated receptors send a visual image to the brain for interpretation. Because the eyes are offset from one another, they return slightly different images back to the brain. An object placed very close to the left eye may stimulate cells in the center of vision in the left eye. The right eye, however, may only have peripheral cells on the edge of the vision stimulated, causing the pencil to appear off to the side. Therefore, the brain would receive two different images: one with the pencil dead center, one with the pencil off to the left. Given the discrepancy in received images, your brain will interpret that the pencil must be very close to you.

The closer an object is to your eyes, the greater the discrepancy in perceived position from one eye to the next. This shift is known as parallax: the difference in apparent position of an object as viewed along two different lines of sight. This same process can be used to determine large distances (such as the distance from Earth to planets, comets, asteroids, and nearby stars) given careful enough measurements and sensitive enough instruments.

With the advent of large telescopes and photography, astronomers were able to systematically photograph and map the entire sky, creating an accurate record of the apparent location of stars and galaxies. Six months later – when Earth had swung around to the opposite side of its orbit and given astronomers their longest possible observational baseline – another all-sky photographic map could be assembled. By layering the two photographs on top of one another, small changes in the apparent locations of a few of the stars became obvious. The changing apparent positions of the stars from one photograph to another were compared and true distances could be computed from simple geometry. Extremely distant stars, clouds of gas, and background galaxies served as the quasi-constant background against which astronomers could detect the measurable parallax angle of much closer stars.

Even relatively close stars (and "close" in terms of stars means several hundred trillion kilometers) display very small shifts, imperceptible to the unaided eye or even small telescopes. The basic unit of stellar distance is the parsec (or pc), the distance at which a star would display a parallax angle of one arcsecond (thus "parsec" is a combination of the words "parallax" and "arcsecond"). The distance to a star is inversely proportional to the parallax angle, meaning that as the distance becomes larger and larger, the parallax angle becomes correspondingly smaller and smaller. With ground based telescopes, astronomers are able to measure with great accuracy the distance to tens of thousands of stars out to a distance of 100 parsecs. The most powerful spaced-based parallax telescope – the ESA's *Hipparcos* telescope – was able to measure stellar distances out to nearly 300 parsecs, a volume encompassing millions of stars.

One of the most important results of parallax measurements is that astronomers could – for the first time – calculate the actual luminosity (light output) of nearby stars. The luminosity of a star is the amount of light emitted from the star's surface and is not a property that can be directly measured. Instead, astronomers must calculate how much light a star produces, which can only be done by measuring a star's distance and the amount of its light reaching the Earth. The further away a light source is, the less light an observer receives. For example, the Sun produces an immense amount of energy. Each square meter of the solar surface produces over 60 million Watts of energy. For perspective, the power output of one 3-foot by 3-foot section of the Sun could power half a dozen large football stadiums. That light radiates from the surface and disperses out into space. By the time that sunlight reaches Earth – at 1 AU – it has been dispersed over a huge area of space. Rather than receiving 60 million Watts per square meter, we only receive 1370 Watts per square meter (enough energy to operate a high power microwave oven). Further from the Sun, at 2 AU, we would receive only one-quarter ($1/2^2 = 1/4$) the amount of sunlight (down to 340 Watts per square meter). Jupiter, which is at approximately 5AU from the Sun, receives one-twenty-fifth the amount of light Earth does ($1/5^2 = 1/25$), which is less energy than is required to operate a standard incandescent light bulb. Distant Pluto, at its furthest point from the Sun (around 40 AU), receives just 1/1600 the Earth's solar radiation.

This relationship between emitted light, received light, and distance is known as the "inverse square law" of light. If astronomers know the distance to a star and the amount of light we are receiving from that star, the inverse square law can be used to calculate how much light the star emits at its surface.

For the first time astronomers could begin to make valid assessments of the nature of stars. Without stellar distances, measuring the actual light output – and thus (indirectly) the forces at work inside of stars, supplying their tremendous energies and slowly changing and evolving them – would have been impossible. Knowing distances, though, allowed astronomers insight into the nature of stars: their energy output, their energy source, their temperature-brightness relationship, the paths of their evolution, and their shape of their eventual deaths.

EQUATIONS AND CONSTANTS

Equation	Expression	Variables
Parallax Equation	$$D = \frac{1}{p}$$	d: the distance in parsec p: the parallax angle in arcsec
Apparent Shift	$\theta = SF \times Ruler$	θ: the apparent shift, in arcsec SF: the scale factor, in arcsec per cm $Ruler$: the physical distance a star moves across a photograph

Continue....

Flux-Distance Equation	$$F = \left(\dfrac{d_{near}}{d_{far}}\right)^2$$	F:	the ratio of flux, or the difference in intensity of light received from two sources
		d_{near}:	the distance from an observer to a nearby light source
		d_{far}:	the distance from an observer to a distant light source
Constants and Conversions			
1 parsec = 3.262 light-years 1 parsec = 206,265 AU			

ILLUSTRATIONS

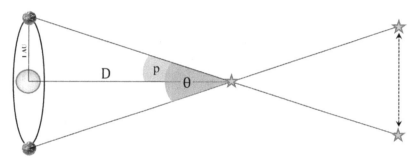

Figure 1. A schematic illustration of the relationship between distance, measured angle, and parallax angle. Two photographs are taken of the star from opposite sides of Earth's orbit. The diagonal lines passing through the star are lines of sight from the Earth to the star. The dashed-line on the far right is the displacement, or the distance the star appears to move across a physical image of the sky. θ is the apparent shift, or the number of arcseconds the star appears to move. The parallax angle – labeled as p – is half of the apparent shift. D represents the distance from the Solar System to the star, and is measured in parsec.

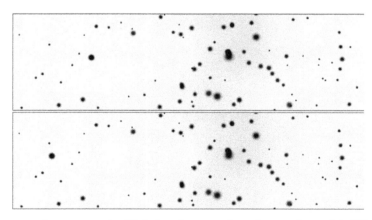

Figure 2. A pair of images of stars near the Pleiades cluster, taken six months and 2 AU apart. One star in the leftmost quarter of the image displays apparent motion due to its close vicinity to Earth (shifting leftward from photo 1 (top) to photo 2 (bottom)). The distance the star appears to travel across the photo is the displacement. Knowing the scale of the photograph (for example, 1 centimeter in the photo may correspond to 0.05" in the sky) allows astronomers to convert the ruler measurement to an angular shift across the sky. The parallax measurement is just half of the angular shift.

PROCEDURE

Datasheets #1 and #2 are photographs of the dense star field in the direction of the Scorpius constellation. In the field of view of the telescope that took the photograph, 1.0 cm of displacement corresponds to 0.025" of angular shift. Datasheet #1's photograph was taken in January, whereas Datasheet #2's photograph was taken in July, six months (and one-half of an Earth orbit) later. The pictures are divided into rows (1 and 2 along the vertical edge of the photograph) and columns (A and B along the horizontal edge of the photograph), separating the photo into quadrants (A1 in the upper left, B1 in the upper right, A2 in the lower left, and B2 in the lower right). Each quadrant contains one star close enough to the Earth to display a measurable parallax over the course of six months. The other stars in the photo are much further away and therefore can serve as an unmoving background.

Flipping back and forth from photo 1 to photo 2, identify the shifting stars. Their motion should be apparent from one observation to the next. Lay a transparency or tracing paper over the first photograph and mark the position of several background stars. Since these distant stars are unmoving, they will serve to accurately align the pictures. Next, mark the position of the four moving stars with a small cross.

Place the transparency on Datasheet #2 and carefully align it using your guide stars. Again, mark the new positions of the four stars with a small cross. By using a ruler, measure the displacements from the Datasheet #1 (January) position to the Datasheet #2 (July) position in centimeters. Record this displacement on Worksheet #1.

Given the scale factor of the photograph, convert the ruler measurement into an apparent shift in arcsec and record that as the angle θ. The parallax angle p is half of the apparent shift. Record the parallax angle in the indicated location on the worksheet.

Use the parallax equation and your data to find the distance to each star in pc; use the conversions to convert this distance into both light-years and AU. Finally, record these calculated values in the spaces provided.

DATASHEET #1

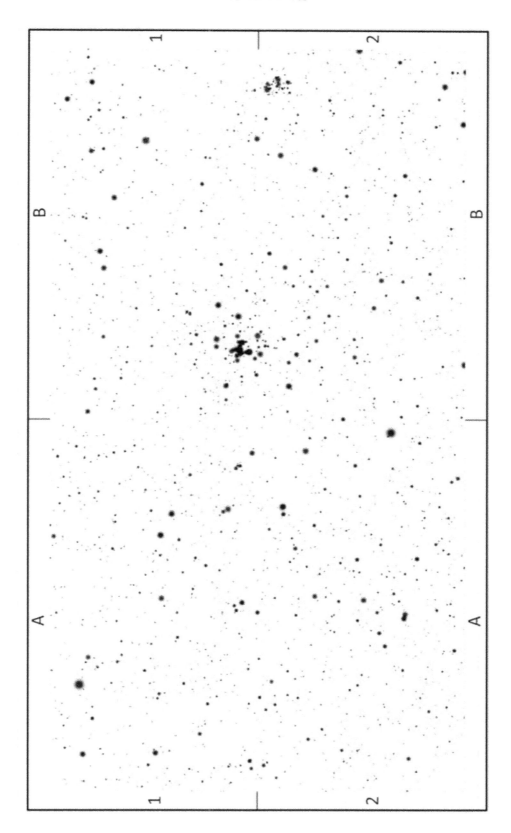

DATASHEET #2

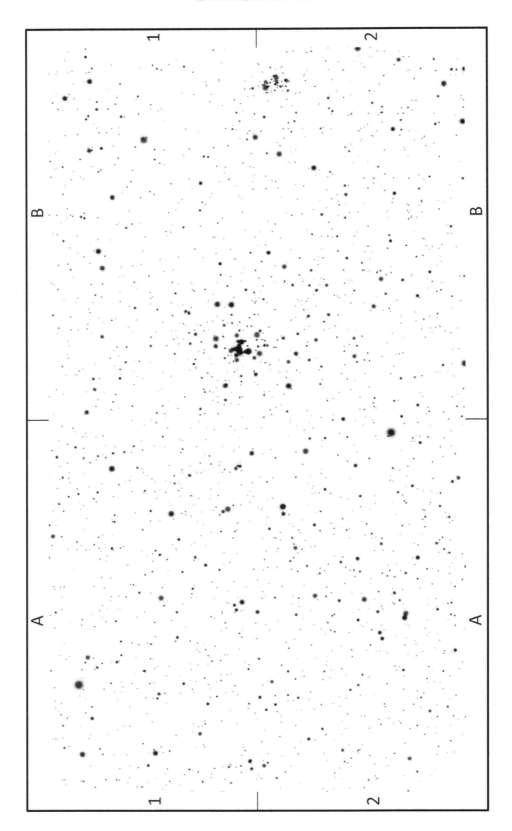

Name _____ Id _____

Due Date _____ Lab Instructor _____ Section _____

Worksheet # 1

Star: HIP 88694	Section A1
Measured Displacement (in cm)	
Apparent Shift (θ in arcsec)	
Parallax Angle (p in arcsec)	
Distance (in pc)	
Distance (in light-years)	
Distance (in AU)	

Star: HIP 88154	Section A2
Measured Displacement (in cm)	
Apparent Shift (θ in arcsec)	
Parallax Angle (p in arcsec)	
Distance (in pc)	
Distance (in light-years)	
Distance (in AU)	

Star: HIP 87472	Section B1
Measured Displacement (in cm)	
Apparent Shift (θ in arcsec)	
Parallax Angle (p in arcsec)	
Distance (in pc)	
Distance (in light-years)	
Distance (in AU)	

Star: HIP 85786	Section B2
Measured Displacement (in cm)	
Apparent Shift (θ in arcsec)	
Parallax Angle (p in arcsec)	
Distance (in pc)	
Distance (in light-years)	
Distance (in AU)	

Name _____ Id _____

Due Date _____   Lab Instructor _____   Section _____

Worksheet # 2

Post-lab questions. Use the equations and conversion factors provided to answer the following questions about the connection between a star's light output, how bright it appears to observers from Earth, and its distance.

The inverse square law of light indicates that the amount of light received from a light emitting object drops rapidly with increasing distance, which is the reason why the stars of the night sky appear only as small points of light and are visible only when the sky is dark. Studies have shown that the star in quadrant A2 (HIP 88154) is nearly identical to the Sun in terms of size, surface temperature, mass, and light output. Therefore, if it was the central star of our Solar System and located 1 AU from planet Earth, the light energy received by us would be the same.

1. If HIP 88154 was located 5.0 AU from Earth (roughly the Sun-Jupiter distance), how much less light would we receive from this star than if it was located at 1 AU?

2. If HIP 88154 was located 30.0 AU from Earth (roughly the Sun-Neptune distance), how much less light would we receive from this star?

3. If HIP 88154 was located 1.0 parsec from Earth (which would make it the closest star to the Solar System), how much less light would we receive from this star?

4. Given the distance you calculated for HIP 88154, what is its brightness compared to the Sun?

5. The star HIP 87472 shines with only 1/33,000,000,000,000th the brightness of the Sun. In reality, is this star more luminous than the Sun or less luminous? In other words: if the Sun was located at the place of HIP 87472, how would the flux or brightness of the Sun and of HIP 87472 compare, according to the inverse square law of light? Show how you came to this conclusion.

UNIT 7.2   THE DIAMETERS AND LUMINOSITIES OF STARS

OBJECTIVE

To learn about the vast size and luminosity differences between various stars located in the Solar System neighborhood. You will be able to draw models of stellar sizes and luminosities to scale to gain insight into their enormous range.

INTRODUCTION

Human scales for measuring distances and brightness – like Watts or kilometers – cannot properly capture the size and scale of many of the Universe's extreme objects. Hence, a scale model helps to envision the proportions of stellar sizes. Reading that the star Betelgeuse, located in the constellation Orion, has a diameter of about 5×10^8 or 500,000,000 km and emits about 16,000 times more energy into space than our Sun does seems meaningless given the staggering scale of these measurements. It is difficult for the mind to comprehend very large (as well as very small) numbers. If a pilot of a jet-liner tells you that you are now flying at an altitude of 37,000 feet, that statement may mean very little in terms of understanding the scope of the plane's height, as it is difficult to envision 37,000 feet. However, if we convert these 37,000 feet into miles, we come up with about 7 mile distance. And that means something to us; it is part of our everyday experience. So, to comprehend the vast scale of the Universe, we must make ourselves some models to the appropriate scale, comparing the diameters of stars to more recognizable objects, such as the diameter of the Earth.

EQUATIONS AND CONSTANTS

Equation	Expression	Variables
Scale Factor Equation	$$Scale = \frac{Ruler}{Real}$$	*Scale*: the scale factor *Real*: the actual size of a real-world object *Ruler*: the scaled down size of the real-world object
Radius	$$R = \frac{D}{2}$$	*R*: the radius of a spherical object *D*: the diameter of that object

PROCEDURE

Worksheet #1, Model 1: This model compares planet Earth to some of the smaller stellar bodies. The 12,700 kilometer wide Earth is drawn as the small circle at the center of the page. Using a ruler, carefully measure the size of the scaled-down Earth. Using the Earth's real and scaled size as the starting point, calculate a scale factor. Using that scale factor, draw and label the following objects:

Barnard's Star

Sirius B - the smallest white dwarf

EG247 - the largest white dwarf

Moon's orbit around the Earth

Sun

Space these object out on the page (do not draw them as concentric circles). Use a compass to smoothly draw the circumference of those objects.

Worksheet #2, Model 2: This model compares the Sun – the largest object from Model 1 – to intermediate sized stars. The 1,500,000 kilometer wide Sun is represented by the small circle at the center of the page. As before, use a ruler to measure the diameter of the scale-model Sun. Using both the Sun's real and scaled size, calculate a new scale factor. Use that scale factor to draw the following properly proportioned objects:

Sirius A

Spica

Capella

Arcturus

Aldebaran

Mercury's orbit around the Sun

Worksheet #3, Model 3: this final model compares the size of our Solar System to some of the galaxy's largest stars. The small centered circle on this worksheet represents the diameter of the giant star Aldebaran (with a diameter 30 × larger than the Sun). Once again, measure the diameter of Aldebaran with a ruler to obtain a ruler measurement. Using the 30 $D_\odot$ measurement for the real diameter of Aldebaran, calculate a new scale factor. Use that scale factor to draw the following properly proportioned objects:

Earth's orbit around the Sun

Mars's orbit around the Sun

Jupiter's orbit around the Sun

Rigel

Antares

Betelgeuse

For the orbits of the planets, draw those as concentric circles centered at Aldebaran to demonstrate the size of planetary orbits in relation to very large stars. Draw Rigel, Antares, and Betelgeuse on their own.

The next portion of the lab – Worksheets #4 through #7 – deals with the light output of these stars. The visible light emitted by a star is known as the luminosity (although, strictly speaking, this term also encompasses light from the entire electromagnetic spectrum), and is normally measured in units of solar luminosities (with the Sun's light output set to 1 $L_\odot$). Much like the physical sizes of the stars vary so greatly, a single on-paper model cannot reasonably represent the luminosity of both the lowest and highest luminosity stars.

Use Worksheet #4 to represent the scaled light output of each of the given stars on three different scales: a scale for low-luminosity stars, where the Sun's light output dominates, a scale for intermediate-luminosity stars, where the Sun is a small fraction of the light output, and a scale for high-luminosity stars, where the Sun's light output in a very small fraction only. The first model compares the luminosity of the Sun to the light output of the following small, low mass, low intensity stars close to the Solar System:

Barnard's Star, Kapteyn's Star, 61 Cygni, ε Eridani, and τ Ceti

The Sun's luminosity serves as the baseline for comparison and is scaled such that 1 $L_{\odot}$ = 100 square centimeters (10 cm across by 10 cm high; this is also equivalent to 10,000 of the small boxes, each of which measures a square millimeter). Worksheet #6 contains a 10 cm by 20 cm graph (in total, there are 200 cm² surface area composed of 20,000 small 1-millimeter by 1-millimeter boxes). The Sun's luminosity is represented by the box labeled "Sun," taking up a full half of the graph. Using the remaining boxes, mark off areas representing the light output of the other 5 stars.

Worksheet #7 compares the light output of the Sun to the moderately bright stars Capella and Sirius A. As on Worksheet #6, the Sun's luminosity is already drawn. However, rather than taking up half of the graph, the Sun's luminosity only occupies a 1-cm by 1-cm square (10-mm by 10-mm). Mark off rectangular areas to represent the light output of Sirius A and Capella.

Finally, Worksheet #8 compares the light output of some of the brightest stars. Now, the Sun's luminosity has been reduced to a single, tiny, 1-mm by 1-mm box. Each one square centimeter box represents 100 $L_{\odot}$. Using this new and final scale, mark off rectangular blocks to represent the light output of the following stars:

Sirius A, Capella, Antares, Betelgeuse, and Rigel

DATASHEET #1

The following table conveys some sizes which will be needed for scaling the size of major stars in Worksheets #1, #2, and #3.

Object	Measurement	
Solar Diameter	1,400,000 km	
Earth Diameter	12,700 km	
Diameter of the Moon's Orbit	770,000 km	
Diameter of Mercury's Orbit	120,000,000 km	0.8 AU
Diameter of Earth's Orbit	300,000,000 km	2.0 AU
Diameter of Mars's Orbit	450,000,000 km	3.0 AU
Diameter of Jupiter's Orbit	1,560,000,000 km	10.4 AU

Star Name	Diameter in Solar Diameters
Largest/Smallest White Dwarf	0.02/0.0067
Barnard's Star	0.2
τ Ceti	0.8
Sirius A	1.8
Spica	7.5
Capella	12
Arcturus	20
Aldebaran	30
Antares	600
Betelgeuse	800

Name _____ Id _____

Due Date _____   Lab Instructor _____   Section _____

Worksheet # 1

Model 1: Sizes of Small Stars

Name _____ Id _____

Due Date _____   Lab Instructor _____   Section _____

Worksheet # 2

Model 2: Sizes of the Sun and Intermediate Stars

Name _____ Id _____

Due Date _____   Lab Instructor _____ Section _____

Worksheet # 3

Model 3: Sizes of Large Stars

Name ______________________ Id ______________________

Due Date __________ Lab Instructor ______________________ Section __________

Worksheet # 4

		Scale Factors				
		Model 5		Model 6	Model 7	
Star Name	**$L/L_\odot$**	**cm^2**	**mm^2**	**cm^2**	**cm^2**	**mm^2**
Sun	1	100	10,000	1	1/100	1
Barnard's Star	0.0005					
Kapteyn's Star	0.004					
61 Cygni	0.08					
ε Eridani	0.3					
τ Ceti	0.5					
Sirius A	24					
Capella	150					
Antares	5,500					
Betelgeuse	13,800					
Rigel	60,000					

1. How many small, millimeter sized boxes are located on the graph on Worksheet #7?

2. How many small, millimeter sized boxes are required to show the light output of the supergiant star Rigel?

3. What percentage of Rigel's total light output will fit on the page on Worksheet #7?

Name _____ Id _____

Due Date _____   Lab Instructor _____   Section _____

Worksheet # 5

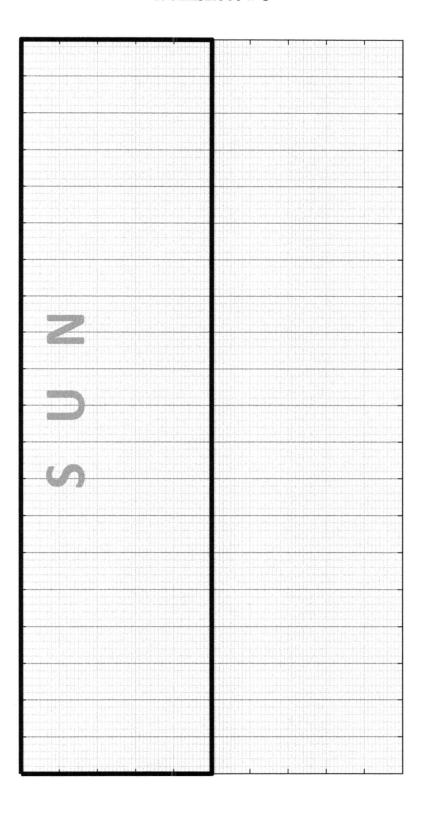

Name _____ Id _____

Due Date _____ Lab Instructor _____ Section _____

Worksheet # 6

Name _____ Id _____

Due Date _____   Lab Instructor _____ Section _____

Worksheet # 7

UNIT 7.3   THE HR AND MK DIAGRAM

OBJECTIVE

To learn how astronomers classify stars using Earth-based observations indicating the stellar temperature. You will be able to use basic information about a star to determine an array of stellar physical properties, such as light output, size, and mass. You will also be able to classify stars based on their evolutionary state.

INTRODUCTION

In human terms, stars are enormously far from the Sun. The closest star to the Solar System – the triple binary Alpha Centauri star system – is over 1.3 pc from Earth (just a little over 4.26 light-years). With present day rocket technology, it would require over 20,000 years to cross this astronomically small distance. From Earth, we can see little more than the light of stars, as even the largest telescopes can barely make out stellar surface features (restricted to nearby stars), such as star spots and prominences. However, there are some stellar characteristics which can be measured and calculated from the Earth, such as the amount of stellar light reaching the Earth, the temperature of the surface gas, and the distance (if the star is sufficiently close).

Telescopes allow astronomers to accurately measure the *apparent magnitude* of a star, i.e., the brightness a star appears to have from Earth, never mind how bright the star actually is. Apparent magnitude can be greatly influenced by distance. A star appearing bright in the night sky does not necessarily mean that the star is actually very luminous and producing large amounts of energy. It may simply be close (as in the case of the Sun). Likewise, a very dim star may in fact be shining with hundreds of thousands of times as much energy as the Sun but may be so distant that barely any of that light reaches the Earth. The *absolute magnitude* is a measure of the brightness a star would have if it was just 10 pc from the Earth. This value can be thought of a measure of the true luminosity of the star.

Astronomers know that a hot object – like a star – has a luminosity that is determined by a combination of the star's temperature and its size. Therefore, a large star will have a higher luminosity than a small star of similar temperature, and a hot star will be brighter than a cool star of the same size. Given how far away most stars are, calculating or measuring a star's size and its luminosity is challenging or even impossible, so while astronomers can determine temperatures of stars, directly measuring the size of the star often remains elusive. Likewise, there is no device that allows astronomers to directly measure or infer a star's luminosity.

However, there are a few thousand stars close enough to the Earth that the parallax distance technique allows astronomers to determine their luminosity, using the property known as the inverse square law. The further away a light source is from an observer, the less of its light will reach the observer. Knowing how much light from a star is reaching the Earth (its apparent magnitude) and how far away a star is, astronomers can work backwards and determine how much light was being produced at the star's surface. Nearby stars range from the small and cool and dim – like Barnard's Star and Wolf 359 – to the hot, the bright, and the large – like Sirius, Procyon, and Mira. Astronomers instantly noticed that the luminosity of a star – the amount of light emerging from its surface – was directly tied to the star's temperature. For nearly every star in the vicinity of the Sun, astronomers could clearly see that the cooler a star was, the dimmer and physically smaller it was. A star slightly hotter than the Sun also has a slightly higher luminosity and usually a slightly larger radius. Stars slightly cooler than the Sun were also slightly dimmer and smaller than the Sun. This pattern held across the entire temperature spectrum.

Therefore, it was entirely possible to simply measure a star's temperature and confidently determine the star's luminosity. The relationship held so well for so many stars that astronomers realized it was entirely possible to tell a star's radius, mass, and luminosity from a study of the star's temperature, called the spectral class. Plotting the temperature versus the luminosity of stars produced a graphical correlation called the main-sequence. Stars were divided into categories called spectral classes according to their similar surface temperature and spectral properties. Spectral classes run from the hottest class – called O stars – to the coolest class – the M stars. The order of the spectral classes, from hottest to coolest is O, B, A, F, G, K, and M. Subclasses more finely divide stars according to their surface temperatures and run from 0 to 9. Therefore, an M0 is hotter than an M1, which in turn has a higher temperature than an M2 and so on. To show that a star is a main-sequence star, astronomers Morgan and Keenan utilized a "luminosity class" designation, written as a "V" after the star's spectral class, thus establishing MK diagrams. The Sun would be considered a G2 V star given its temperature and luminosity.

As with all populations, there are always outliers: individuals that don't follow expected or established patterns. While it was nearly always true that the coolest stars were the dimmest stars, some extremely cool stars – like Betelgeuse, Antares, and Aldebaran – actually turned out to be terrifically bright (more than 50,000 $L_\odot$ for Betelgeuse, 10,000 $L_\odot$ for Antares, and 450 $L_\odot$ for Aldebaran). Astronomers were puzzled by these unexpectedly and abnormally luminous, low temperature stars. Astronomers studying these stars realized their tremendous luminosities came from their huge diameters rather than a high surface temperature. While they weren't extremely hot, their gargantuan size – upward of 1000 $R_\odot$ (early estimated value) for Betelgeuse, or 1000× the diameter of the Sun – supplied a huge area for radiating photons. With so much surface area, even low-temperature gas could produce a flood of photons, causing the star to be tremendously luminous. Obviously, these stars could not be considered main-sequence (V luminosity class) stars as they did not follow the main-sequence trend. New luminosity classes were established dependent on the star's light-output and size. These classes were called "red giant" (designated by a III), "supergiants" (II), and "hypergiants" or "bright supergiants" (I). Stars belonging to these types have high light outputs that can be traced to the large stellar sizes.

Modern understanding of stars and their evolution establishes that these luminosity classes are populated by stars which are nearing the end of their stellar lives. As stars deplete their internal hydrogen fuel reserves and begin to succumb to strong internal gravitational forces, their fusion rates increase dramatically. The increased energy output from the core propagates through the upper stellar layers and heats the outer envelope, causing it to expand outward. Eventually, when these stars have exhausted all of their available fuel, their stellar lives will end, sometimes in the gentle ejection of those heated, puffed up outer layers, and sometimes in the destructive, explosive collapse of the star's core. The stars which end their lives by releasing their outer layers into space leave behind a very dense, very hot stellar remnant called a white dwarf, designated by the luminosity class "wd." A white dwarf is the carbon-oxygen rich core of the former star, depleted of useable nuclear fuel and exposed when giant star's outer layers were shrugged off into space as a planetary nebula. These cores, while hot, are extremely small (usually akin to the size of the Earth) and dim owing to their small sizes.

The Hertzsprung-Russell (HR) diagram depicts a compilation of stars charting their absolute magnitudes versus observed surface temperatures. It also shows the various placements of the various classes of stars: the orderly main-sequence; cool, bright red giants; the dim, hot white dwarfs; the supergiants which are extremely bright no matter what their surface temperature is.

EQUATIONS AND CONSTANTS

Equation	Expression	Variables	
Magnitude-Luminosity	$L = 2.512^{M_{V\odot} - M_V}$	L:	the luminosity of a star in $L_\odot$
		$M_{V\odot}$:	the absolute magnitude of the Sun
		M_V:	the absolute magnitude of a star
Distance Modulus	$\mu = m_V - M_V$	m_V:	the star's apparent magnitude
		M_V:	the star's absolute magnitude
Distance	$D = 10^{(\mu + 5)/5}$	D:	distance in pc
		μ:	the distance modulus

Constants and Conversions
$M_{V\odot} = 4.83$
1 parsec = 3.262 light-years

ILLUSTRATIONS

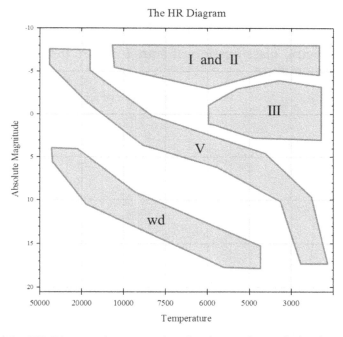

The HR Diagram

Figure 1. A schematic of the HR Diagram demonstrating the clustered correlation between temperature and light output for stars including the hydrogen-fusing main-sequence (luminosity class V), the evolved, hydrogen-exhausted giants (III), and the highly luminous, extremely large supergiants, bright supergiants, and hypergiants (II and I). Notice that the class of subgiants (luminosity class IV) has been omitted for editorial reasons. Those stars are located slightly above the main-sequence.

PROCEDURE

Worksheet #1 contains two tables: one with the 5 brightest stars in the sky, plus the Sun as an example, and one with the 5 closest stars. Use Tables 4 and 5 of the Appendix to record the spectral class, absolute magnitude, and luminosity class of those 11 stars. Using the absolute magnitudes, calculate the luminosity of each star in units of solar luminosities. For each star, use the graph on Datasheet #1 to determine the surface temperature – in Kelvin – for each of the stars. For the main-sequence stars, also determine their diameters and masses. For stars whose masses cannot be determined due to their advanced state of evolution, simply cross out that box or write "N/A" in the space provided.

Worksheet #2 lists 11 stars. Plot each of the stars in the HR diagram provided on Worksheet #4. From the star's location on the graph, determine its luminosity class and record it. Using the magnitude-luminosity equation, calculate the light output of each star in units of solar luminosities $L_\odot$ and use Datasheet #1 to determine each star's surface temperature.

Worksheet #3 contains a list of stars known to host Earth-type planets (in the broad sense of this term). Given the distance to these stars, astronomers are only able to reliably measure the stellar surface temperatures. Since they are main-sequence stars, their temperatures correlate closely to their masses and diameters. Determine their absolute magnitude through placing the stars in their expected main-sequence position in the HR diagram.

Worksheets #5 and #6 are provided as supplements. Worksheet #5 has spaces provided for listing and calculating the properties of up to 25 stars. Using either a datasheet or an online data source like the SIMBAD database, fill in the datasheet with pertinent information, using the datasheet temperature-spectral class and (for main-sequence stars) mass/diameter-spectral class relations to determine the stellar properties. Worksheet #6 contains an empty HR diagram which may be used for the plotting a selection of stars and the construction of the HR diagram for the stellar data.

DATASHEET #1

This graph matches the spectral classification of stars for stars between spectral types B0 and M8 to the stellar surface temperature.

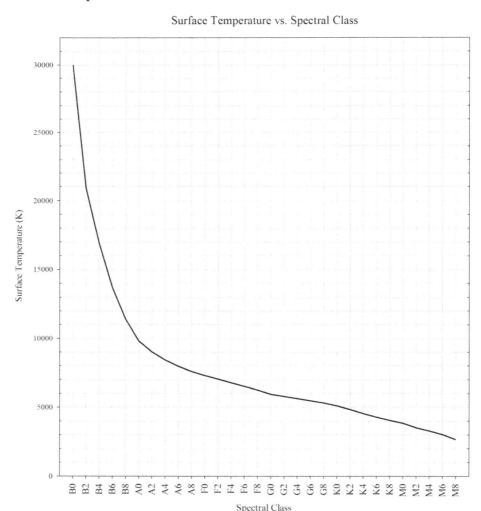
Surface Temperature vs. Spectral Class

DATASHEET #2

For main-sequence stars (luminosity class V), diameters and masses are strongly correlated to the stellar surface temperatures. The graphs below indicate the correlations for masses and diameters regarding main-sequence stars between spectral types B0 and M8.

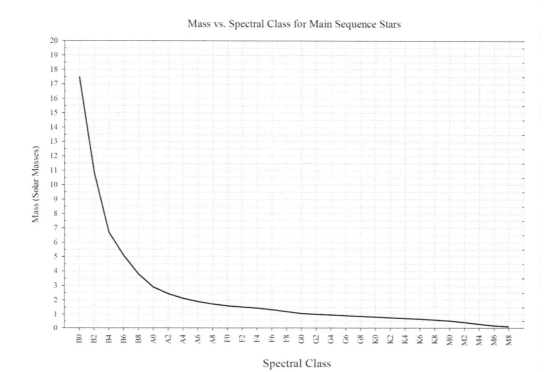

Mass vs. Spectral Class for Main Sequence Stars

Diameter vs. Spectral Class for Main Sequence Stars

Name _____ Id _____

Due Date _____   Lab Instructor _____   Section _____

Worksheet # 1

The 5 brightest stars in the sky plus the Sun:

Star	Spectral Class	M_V	Luminosity (in $L_\odot$)	Temperature (in K)	Diameter (in $D_\odot$)	Mass (in $M_\odot$)	Luminosity Class
Sun	G2	4.83	1.00	5800	1.00	1.00	V
Sirius							
Canopus							
Arcturus							
Vega							
Capella							

The 5 nearest stars to the Solar System:

Star	Spectral Class	M_V	Luminosity (in $L_\odot$)	Temperature (in K)	Diameter (in $D_\odot$)	Mass (in $M_\odot$)	Luminosity Class
Proxima Centauri							
Alpha Centauri A							
Alpha Centauri B							
Barnard's Star							
Wolf 359							

Name _____ Id _____

Due Date _____   Lab Instructor _____   Section _____

Worksheet # 2

A Selection of Stars:

Star	Spectral Class	M_V	m_V	Luminosity (in $L_\odot$)	Distance (in $D_\odot$)	Luminosity Class
Porrima	F0	3.04	3.40			
IX Carina	M2	−2.93	7.75			
Mirphak	F5	−4.21	1.75			
Zeta Reticuli	G2	5.10	5.50			
YY Geminorum	M1	8.96	9.95			
Keid	K1	5.92	4.43			
Zubenelgenubi	A3	0.92	2.75			
Alfirk	B2	−3.41	3.20			
Deneb	A2	−6.93	1.25			
40 Eridani B	B4	11.01	9.52			
Sadalmelik	G2	−3.08	2.96			

1. Many stars in the sky are binaries, meaning that two stars formed from the same cloud of gas and orbit one another. Which stars above are paired together in a binary? What tells you that these pairs of stars are likely physically bound to one another?

2. Which star has the highest surface temperature? The lowest?

3. Which star shares the most similar characteristics with the Sun?

4. Which star is most distant? How far away is it in light years? What year did the light that we see from this star leave the star's surface?

Name _____ Id _____

Due Date _____ Lab Instructor _____ Section _____

Worksheet # 3

The following main-sequence stars are known to host Earth-type exoplanets (in the broad sense of this term), though their distances only allow astronomers to reliably measure the stellar surface temperatures. Using the HR diagram and data sheets, determine the major physical properties of these stars.

Name	Temperature (in K)	Spectral Class	Absolute Magnitude	Diameter (in $D_\odot$)	Mass (in $M_\odot$)
Kepler 86	5620				
55 Cancri	5370				
Kepler 186	3790				
Kepler 62	4900				

Name _____ Id _____

Due Date _____   Lab Instructor _____   Section _____

Worksheet # 4

The HR Diagram

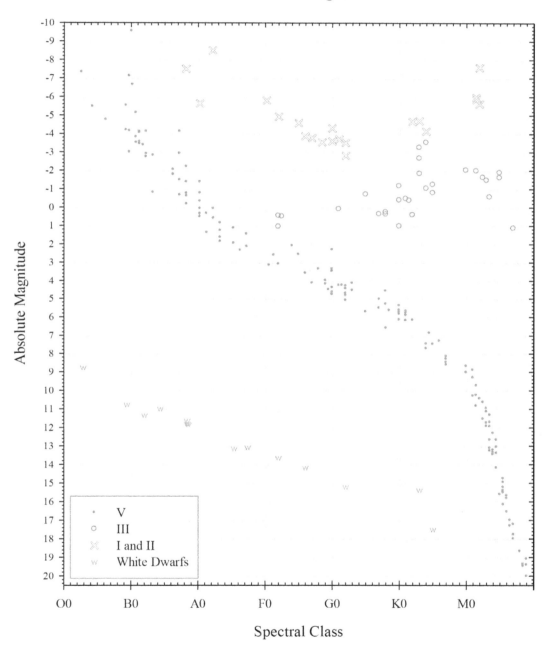

Name _____ Id _____

Due Date _____ Lab Instructor _____ Section _____

Worksheet # 5

Star	Spectral Class	M_V	Luminosity (in $L_\odot$)	Temperature (in K)	Diameter (in $D_\odot$)	Mass (in $M_\odot$)	Luminosity Class

Continue....

Star	Spectral Class	M_V	Luminosity (in $L_\odot$)	Temperature (in K)	Diameter (in $D_\odot$)	Mass (in $M_\odot$)	Luminosity Class

Name _____ Id _____

Due Date _____   Lab Instructor _____   Section _____

Worksheet # 6

The HR Diagram

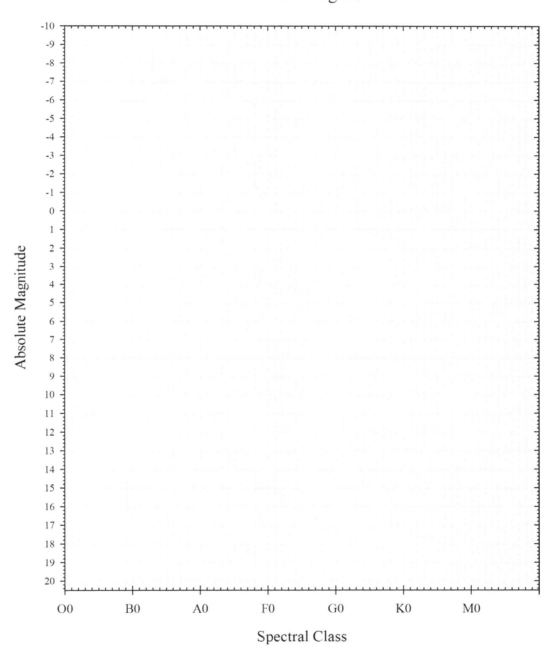

Spectral Class

UNIT 7.4   CEPHEIDS AND OTHER PULSATORS

OBJECTIVE

To learn how the steadily changing light patterns of unique, evolved stars gave astronomers the ability to stretch distance measurements beyond the Milky Way to galaxies hundreds of millions of light-years away. You will be able to plot the changing light levels of pulsating stars and use that data to find luminosities and distances of extragalactic objects.

INTRODUCTION

The beginning of the 20th century saw astronomers debate the nature, size, and origin of the Universe as ever more powerful telescopes allowed for ever deeper and more precise observations. One looming question centered on the size of the Universe; namely, was the Milky Way galaxy the entire Universe or was it simply one of a huge number of bodies inhabiting a vast universe. At the heart of this debate were mysterious, poorly understood objects called "spiral nebulae," so named because they appeared like fuzzy, cloudy, non-descript blobs with glowing centers and vaguely spiral shapes. Through large telescopes using sensitive photographic techniques available at the time, astronomers were unable to classify the spiral nebulae. Some believed that they were common gas clouds, like the hundreds of nearby nebula identified in the disk of the Milky Way, whereas others postulated that they were "island Universes": huge conglomerations of dust and gas like the Milky Way itself.

Part of the problem with determining their nature was calculating their distances. Not surprisingly, they were beyond the reach of geometric parallax, meaning that the spiral nebulae were at least 100 pc distant. More troubling, individual stars within the nebula could not be identified, making main-sequence magnitude fitting impossible.

The star RR Lyrae – located just over 260 pc from the Earth – was discovered to have a unique property. Studies of its temperature and brightness showed that it varied rapidly over the course of just over 12 hours. The star's brightness rose and felt by 50% as its surface temperature swung between 8000 K and 6300 K. RR Lyrae was just the first of a new type of ancient, pulsating star type to be discovered: the RR Lyra variables. These stars are low mass, highly evolved, unstable giant stars which undergo rapid oscillations in light output. While not highly luminous, they are quite uniform in the pulsation periods – generally 0.5 days – and luminosity – with absolute magnitudes of 0.75; their luminosities correspond to approximately 45× that of the Sun. Because of their rapid and regular pulsations – as well as their uniform absolute magnitudes – RR Lyrae variables are easy to spot in a crowded star field and became highly important for measuring the distance to distant star clusters. RR Lyrae stars are predominantly found amongst the ancient stellar populations of the Milky Way's globular clusters.

A second major class of variable stars – the Cepheids – is named after the nearby star Delta Cephei. As early as the 1700s, astronomers realized that Delta Cephei changes in brightness with a very regular 5.37-day period. Beginning at maximum brightness, the star would fade steadily over the course of the next four days, dimming to under half its brightness before quickly ramping back to its maximum light output. Later studies showed that this cycling brightness was due to Delta Cephei expanding and contracting like a beating heart. It was also found that Delta Cephei was extremely luminous, far in excess of the moderately luminous RR Lyrae stars. With an absolute magnitude of –3.47, this pulsating star is emitting over 2000× the light output of the Sun, i.e., 50× the luminosity of RR Lyrae. As astronomers identified other pulsators similar to Delta Cephei – stars like Eta Aquilae, S Sagittae, and RS Puppis – it became clear that Cepheids demonstrate a remarkable and unique trend: the longer the time required to complete one pulsation, the higher the star's luminosity. This particular type of pulsating stars with long periods and high luminosities were classified as Cepheids.

Delta Cephei itself proved to be a low-luminosity Cepheid, despite its 2000 $L_\odot$ light output. Longer-period Cepheids could reach absolute magnitudes of −6 and higher, corresponding to luminosities in excess of 20,000 $L_\odot$, 500× brighter than the common RR Lyrae stars. Cepheids themselves fall into two categories: Type I and Type II. Type II Cepheids are cool, ancient, lower mass supergiant stars. They are markedly less luminous than their Type I counterparts. The Type I Cepheids are much younger, hotter, and more massive, shining at substantially higher luminosities for similar pulsation periods.

Armed with these extremely luminous stars, astronomers were able to determine distances to these faint "spiral nebulae." For the first time it was shown conclusively that these faint smudges of light were actual whole galaxies far outside of the realm of the Milky Way. While the Milky Way galaxy was calculated to extend some 30 kpc in diameter, some of these Cepheid stars were calculated to lie in spiral nebulae that were millions of kpc away from the Solar System. Cosmology – the study of the Universe in its entirety – was born.

EQUATIONS AND CONSTANTS

Equation	Expression	Variables
Distance Modulus	$\mu = m_V - M_V$	m_V: the apparent magnitude M_V: the absolute magnitude
Distance-Magnitude Relation	$D = 10^{(\mu+5)/5}$	D: the distance in parsecs μ: the distance modulus
Luminosity	$L = 2.512^{4.83-M_V}$	L: luminosity in solar luminosities M_V: absolute magnitude of the star
Constants and Conversions		
1 kpc = 1,000 pc		
1 Mpc = 1,000 kpc = 1,000,000 pc		
$M_{V\text{-RR Lyrae}} = 0.75$		

ILLUSTRATIONS

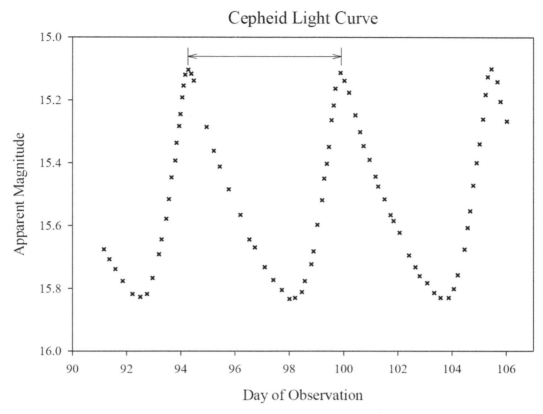

Figure 1. A typical light curve showing the changing apparent magnitude of a Cepheid during variation. The magnitude of this star varies between $m_V = 15.1$ and 15.9. The peak apparent magnitude occurs on day 94.25 at an apparent magnitude of 15.10. The next peak occurs at $m_V = 15.11$ on day 99.88. The period of this Cepheid is 5.63 days, which indicates an absolute magnitude of –3.4 for a Type I Cepheid. Using the distance equation along with the data ($m_V = 15.10$ and $M_V = -3.4$), astronomers were able to conclude that this Cepheid is at a distance of 50 kpc.

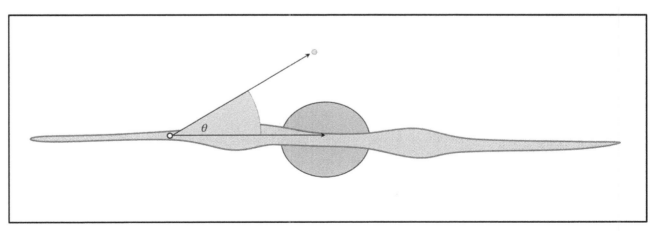

Figure 2. A simple schematic of the Milky Way galaxy showing the central bulge and the thin, extended disk (note that the disk's width is exaggerated here). The Sun's location in the disk is marked with a white circle. The Milky Way is surrounded by a halo of orbiting globular clusters as well as a more distant assortment of small "dwarf galaxies" such as the Ursa Minor dwarf and the Leo I dwarf. The position of these objects can be pinpointed in relation to the Milky Way using two measurements: its distance from the Earth and its direction with reference to the center of the Milky Way. 0° galactic latitude points directly toward the center of the Milky Way's nucleus. 90° points directly up above the Sun. 180° points directly away from the nucleus (out through the edge of the disk). 270° points directly below the Sun's position. The object shown in the drawing is located at an angle of 30° above the disk and at a distance of 13 kpc.

PROCEDURE

Datasheet #1 demonstrates the period-luminosity relationship for both Type I and Type II Cepheids (Type I Cepheids are intrinsically brighter than Type II Cepheids by their stellar nature, though both types far outshine main-sequence stars like the Sun). Refer to this relationship to determine the true light output (absolute magnitude) for Cepheids of a given period.

For Worksheet #1, you will plot the typical light curve of a Cepheid pulsator. Data from a telescope is available in Datasheet #2, where the incoming light level (apparent magnitude) of the star was measured at regular intervals for a span of three weeks while the star brightened and dimmed. Plot the data on the worksheet provided and connect the data points with a smooth, best-fit curve.

Worksheet #2 demonstrates how astronomers were able to use variable stars to determine the distance to a number of small, orbiting galaxies located in the vicinity of the Milky Way galaxy. Much like large planets are orbited by various moons, the massive Milky Way is located within a swarm of small, low-mass, irregular "dwarf" galaxies. Using your knowledge of variable stars and their absolute magnitudes, calculate the distances to these small galaxies based on the type of variable star found within them.

The graph on Worksheet #3 is centered on the Sun's galactic position, with the Milky Way galaxy's nucleus and disk shown in light gray. The concentric circles on the graph each indicate 20 kpc distances from the Sun. Using the angular positions given for each of the Milky Way's dwarf galaxies (θ) and the distance determined from Worksheet #2, plot the positions of the Milky Way's satellite galaxies. For example, if a dwarf galaxy was located at 110 kpc at an angular position of 90°, you would find 90° on the graph (which would be a position directly above the Sun, perpendicular to the disk of the galaxy) and measure a distance out to the 110 kpc distance. Draw a small mark at that position and label the dwarf galaxy's name.

DATASHEET #1

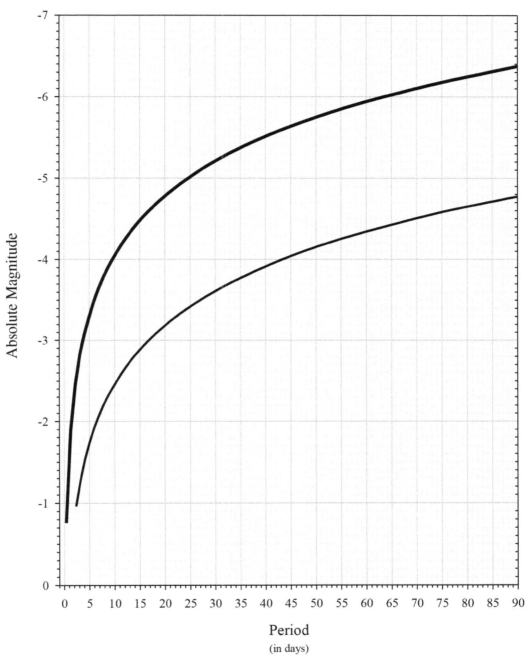

Cepheid Period-Luminosity Relationship

DATASHEET #2

Observations for the Cepheid pulsator XX Persei

Observation Number	Day	Apparent Magnitude
1	22.5523	14.8226
2	22.9111	14.3790
3	23.2340	13.7016
4	23.6287	12.9919
5	23.9516	12.7500
6	24.0952	12.7016
7	24.3822	12.7339
8	24.7769	12.8629
9	25.5663	13.1855
10	26.2839	13.4919
11	27.0733	13.9194
12	28.0062	14.2742
13	28.5445	14.4194
14	29.4774	14.6290
15	31.5944	15.0968
16	32.9579	15.3145
17	34.3573	15.4919
18	35.8643	15.5887
19	36.7254	15.6129
20	37.3354	15.5645
21	38.0172	15.2742
22	38.9142	14.3226
23	39.5242	13.0645
24	39.7753	12.7661
25	40.1342	12.7016
26	41.0671	12.9516
27	42.1435	13.3952

Name _____ Id _____

Due Date _____   Lab Instructor _____   Section _____

Worksheet # 1

Light Curve of XX Persei

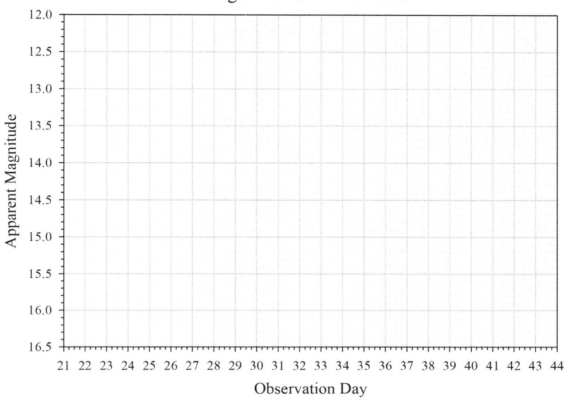

Observation Day

Period (in days)	
Maximum Apparent Magnitude	
Absolute Magnitude (for a Type II)	
Absolute Magnitude (for a Type I)	

1. If astronomers concluded that XX Persei was a Type II Cepheid (ancient, moderate temperature, and located in a globular cluster), what would be the calculated distance to this star?

2. If XX Persei was determined to be a Type I Cepheid (young, high temperature, and located in a galactic spiral arm), what would be the calculated distance to this star?

Name _____ Id _____

Due Date _____ Lab Instructor _____ Section _____

Worksheet # 2

Large Magellanic Cloud

Location: 32° below the disk of the Milky Way   (θ = 212°)

Star Type	Period (in days)	Apparent Magnitude (m_V)	Absolute Magnitude (M_V)	Distance (in kpc)
RR Lyrae	0.534	19.24		

Small Magellanic Cloud

Location: 44° below the disk of the Milky Way   (θ = 224°)

Star Type	Period (in days)	Apparent Magnitude (m_V)	Absolute Magnitude (M_V)	Distance (in kpc)
Type I	12.22	14.63		

Carina Dwarf Spheroidal Galaxy

Location: 22° below the disk of the Milky Way   (θ = 202°)

Star Type	Period (in days)	Apparent Magnitude (m_V)	Absolute Magnitude (M_V)	Distance (in kpc)
Type II	8.25	17.75		

Draco Dwarf Spheroidal Galaxy

Location: 34° above the disk of the Milky Way   (θ = 34°)

Star Type	Period (in days)	Apparent Magnitude (m_V)	Absolute Magnitude (M_V)	Distance (in kpc)
Type I	25.00	14.51		

Bootes I Dwarf Galaxy

Location: 69° above the disk of the Milky Way   (θ = 69°)

Star Type	Period (in days)	Apparent Magnitude (m_V)	Absolute Magnitude (M_V)	Distance (in kpc)
RR Lyrae	0.555	19.64		

Continue....

Sculptor Dwarf Galaxy

Location: 83° below the disk of the Milky Way   ($\theta = 277°$)

Star Type	Period (in days)	Apparent Magnitude (m_V)	Absolute Magnitude (M_V)	Distance (in kpc)
Type II	36.08	15.97		

Fornax Dwarf Spheroidal Galaxy

Location: 65° below the disk of the Milky Way   ($\theta = 245°$)

Star Type	Period (in days)	Apparent Magnitude (m_V)	Absolute Magnitude (M_V)	Distance (in kpc)
Type I	64.10	14.73		

Ursa Minor Dwarf Spheroidal Galaxy

Location: 44° above the disk of the Milky Way   ($\theta = 44°$)

Star Type	Period (in days)	Apparent Magnitude (m_V)	Absolute Magnitude (M_V)	Distance (in kpc)
Type I	2.56	16.29		

Sextans Dwarf Spheroidal Galaxy

Location: 42° above the disk of the Milky Way   ($\theta = 138°$)

Star Type	Period (in days)	Apparent Magnitude (m_V)	Absolute Magnitude (M_V)	Distance (in kpc)
Type II	12.22	17.11		

Sagittarius Dwarf Spheroidal Galaxy

Location: 14° below the disk of the Milky Way   ($\theta = 346°$)

Star Type	Period (in days)	Apparent Magnitude (m_V)	Absolute Magnitude (M_V)	Distance (in kpc)
RR Lyrae	0.496	17.25		

Name _____ Id _____

Due Date _____   Lab Instructor _____   Section _____

Worksheet # 3

The Vicinity of the Milky Way

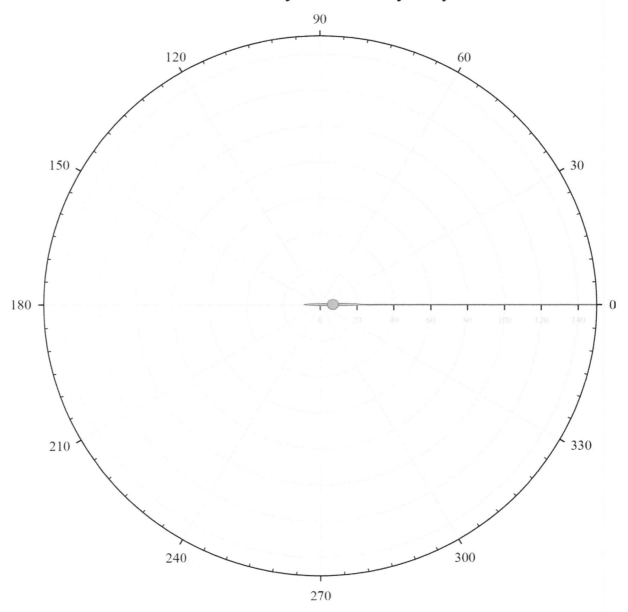

Name _____ Id _____

Due Date _____   Lab Instructor _____   Section _____

Worksheet # 4

Post lab questions: Refer to the constants and equation sheets to answer these questions. Show all work and equations used.

1. The Andromeda galaxy – the largest nearby "spiral nebula" known to astronomers – is also the closest major galaxy to the Milky Way. A very luminous, 85-day Type I Cepheid was found in this galaxy to have an apparent magnitude of only 18.15 (appearing nearly 100× dimmer than the dwarf planet Pluto through a telescope). How far away is the Andromeda galaxy, in kpc?

2. The Milky Way galaxy's spiral disk is 30 kpc across. Laid out end to end, how many Milky Way galaxies would fit into the space between the actual Milky Way galaxy and the Andromeda galaxy?

3. The brightest Type I Cepheid has an absolute magnitude of –6.6, while the dimmest object that can be reliably detected by modern day telescopes has an apparent magnitude of 28.0 (although this number is expected to change due to technological progress). How far away could the most luminous Cepheid be before even the largest telescope could no longer detect its light? In other words: How far from the Milky Way can the Cepheid distance determination method be used before even these luminous stars are so dim and distant that they are invisible?

UNIT 7.5   GLOBULAR CLUSTERS AND MOLECULAR CLOUDS

OBJECTIVE

To learn how astronomers' observations of the galaxy's large, orbiting gas clouds and the distant galactic star clusters led to our understanding of the scale and extent of the Milky Way galaxy

INTRODUCTION

From our Earth-bound perspective, the Milky Way galaxy looks like little more than a dim swath of light cutting north-to-south through the sky. Ancient astronomers used mythological stories to describe the origins of the Milky Way, concluding that it was milk spilled across the sky by a nursing Zeus. It was not until the 1600's and the advent of the telescope that Galileo Galilei could demonstrate that the dim, murky light of the Milky Way's prominent band was actually the combined light of many thousands of tightly packed stars, too dim, distant, and close to one another for the human eye to distinguish them; thus, they appeared merely as a blur of light. Later astronomer William Herschel attempted to gauge the size and shape of the Milky Way galaxy using star counts. His telescopic survey work centered on studying portions of the sky in different directions and counting the number of visible stars. He found that the number of stars dropped off evenly in almost every direction, leading him to conclude that the galaxy was flat, wide, moderately large at approximately 1000 pc, and the Sun and Solar System were located very near to the center. While incorrect in most aspects owing to severe limitations in his equipment and understanding of stellar luminosity, his model was a serious early scientific attempt to draw an accurate scale model of the Galaxy.

The astronomer Charles Messier used his telescope to begin identifying unique objects in the sky, including planetary nebulae (which are actually the remains of extinct, Sun-like stars which have recently ended their stellar lives), "spiral nebulae" (dim, vaguely spiral wisps of light which actually turned out to be very distant galaxies independent of the Milky Way galaxy), open star clusters, globular clusters, and nebulae (large clouds of galactic hydrogen and helium gas). Generation by generation, astronomers were forming a more complete picture of the make-up and population of the Milky Way galaxy, even if the size was still proving elusive.

Astronomy at the turn of the 20th century featured intense debate about the nature of the Milky Way galaxy: its size, its extent, and its place in the Universe at large. It was even questioned whether or not the Milky Way itself *was* the entire Universe. At issue was the inability to clearly see any center, border or major coherent structures in the Milky Way. This is largely caused by our location. The Solar System is located deep within the disk of the Milky Way and surrounding clouds of dust and gas obscure our view of the surrounding galactic environment. In particular, a large, dark molecular cloud (a dense, cold pillar of molecular hydrogen gas) obscures our view of the Milky Way's center region, making distance determination incredibly difficult.

Messier's catalogue of unique objects would play a key role in determining the size, shape, and mass of the Milky Way galaxy. In particular the globular clusters – large, coherent, intrinsically luminous balls of ancient stars – were seen as a way of inferring the position of the Milky Way's center. Populated with 1000s to 100,000s of stars, the combined light of a globular cluster may shine as bright as a supergiant star. Furthermore, many globular clusters are located above or below the plane of the Milky Way, thus appearing highly detached from the obscuring dust and gas clouds, which considerably hinder astronomer's view through the Milky Way's disk.

Finally, globular clusters are hosts to bright, predictably luminous RR Lyrae and Cepheid stars, making their distance determination reliable and accurate. Armed with known distances and locations in the sky, a map of the Milky Way's 150 globular clusters could be created. Astronomer Harlow Shapley assumed that the globular clusters orbit the Milky Way in much the same way planets orbit a star or moons orbit a planet. Furthermore, he argued, just as the Sun sits at the center of mass of its planets' orbits and just as a planet sits at the center of its moons' orbits, the orbit of globular clusters should be centered on the Milky Way galaxy's center.

With the advent of radio and infrared telescopes, the distances and locations of the Milky Way's dark nebula – the cold, large, dense molecular clouds – could be pinpointed as well, allowing astronomers to begin tracing out the shape of the Milky Way's disk. In particular, the Milky Way's gas clouds traced out a concentrated spiral arm pattern, with the large gas clouds aligning into sweeping, patchy columns winding around the Galactic core.

EQUATIONS AND CONSTANTS

Equation	Expression	Variables
Circumference	$C = 2\pi R$	C: the length of a circular orbit R: the orbital radius (distance from a point on the orbit to the orbit's center)
Period-Distance-Velocity Relationship	$P = \dfrac{C}{v}$	P: the time required to complete one orbit v: the velocity of the orbiting body
Kelper's Third Law	$M = \dfrac{a^3}{P^2}$	M: the mass of the central object (in solar masses) a: the distance between an orbit and its center (in AU) P: the time required to complete one orbit (in years)
Constants and Conversions		
1 kpc = 1,000 pc = 2.06265×10^8 AU		
1 year = 3.15576×10^7 sec		

ILLUSTRATIONS

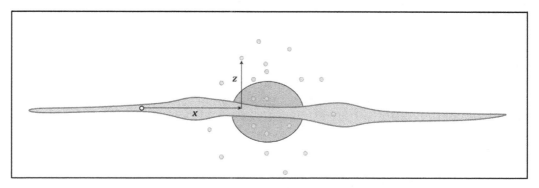

Figure 1. A simple edge-on schematic of the Milky Way galaxy showing the central bulge and the thin, extended disk (its width is exaggerated here). The Sun is shown in its generally proper location, "submerged" in the material making up the thin disk. The material of the disk makes the direct locating of the galaxy's center extremely difficult. But by measuring the distance to and position of globular clusters, the position of the center can be inferred. The plotting of a globular cluster is shown with the x-coordinate denoting its distance along the disk and its z-coordinate denoting its distance above or below the disk.

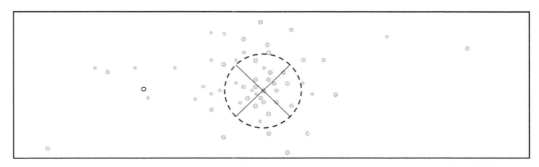

Figure 2: Harlow Shapley showed that the center of the concentration of globular clusters also pointed to the center of the mass of the Milky Way. Drawing a circle around the heaviest concentration of globular clusters and pinpointing the center of that concentration allows astronomers to determine the position of – and distance to – the Milky Way's center of mass in an approximate manner.

PROCEDURE

Datasheet #1 contains a large number of data points, corresponding to the positions of the Milky Way's 150+ globular clusters. The positions are given in x- and z-coordinates: distances along the disk and distances above or below the disk are in kpc. Using either the graph paper provided or a graphing program, plot the positions of the globular clusters. The boundaries of the graph should be set at −50 kpc to +50 kps along the horizontal axis (which points along the length of the Milky Way's disk) and −30 kpc to +30 kpc along the vertical axis (which points above and below the Milky Way's disk). Since these observations were made from the position of the Sun, the Sun will be centered at the point (0,0).

For far-outlying globular clusters beyond the borders of the graph, do not graph those positions. An x-coordinate beyond the boundaries of +50 / −50 kpc or a z-coordinate beyond the boundaries of +30 / −30 kpc corresponds to a far-flung cluster, which will not influence the estimate of the galaxy's center. Many of the globular clusters will obviously be located in a somewhat compacted, concentrated region of the graph.

Once the positions have been graphed, estimate the center of the concentration. Notice that there are generally as many clusters above the Sun's position (positive z) as there are stars below the Sun's position (negative z values). This verifies that the Solar System is located in the middle of the Milky Way's disk and that a line running through the edge of the disk pointing to the galactic core will pass through the Solar System.

Datasheet #2 contains a large number of molecular gas clouds' positional data. The x-coordinate data runs across the disk of the Milky Way with the y-coordinate data pointing toward the galactic center. Using either the graph paper provided or a graphing program, plot the positions of the globular clusters. The boundaries of the graph should be set at -20 kpc to $+20$ kps along the horizontal axis (which points along the length of the Milky Way's disk) and -20 kpc to $+20$ kpc along the vertical axis (which points above and below the Milky Way's disk). Since these observations were made from the position of the Sun, the Sun will be centered at the point (0,0).

Once the positions have been graphed, notice concentrated arcs highlighted by the star-forming gas clouds. The arms of spiral galaxies tend to wrap symmetrically around the central nucleus of the galaxy. Lightly complete the arc of the visible arms and estimate the position of the center of the galaxy by estimating the center of the arms' winding.

Using either graph, you should be able to estimate the position of the center of the Milky Way galaxy (in units of kpc). Once the size of the galaxy is known, its mass can be estimated from Kepler's third law, which relates periods, distances, and masses.

DATASHEET #1

GLOBULAR CLUSTERS

#	Name	X	Z	#	Name	X	Z
1	1636−283	8.03	1.74	16	IC 4499	23.13	6.53
2	2MASS−GC01	3.54	0.01	17	Kosopov 1	4.99	0.53
3	2MASS−GC02	4.83	−0.05	18	Kosopov 2	10.69	−6.57
4	AM 4	20.65	17.78	19	Liller 1	−2.49	45.6
5	Arp 2	26.44	10.15	20	NGC 104	1.87	−3.18
6	BH 176	16.05	1.43	21	NGC 1261	0.09	−12.87
7	BH 261	6.46	−0.6	22	NGC 1851	−4.26	−6.95
8	Djorg 1	13.66	0.59	23	NGC 2298	−4.28	−2.98
9	Djorg 2	6.29	0.27	24	NGC 2808	1.99	−1.87
10	E 3	2.9	−2.64	25	NGC 288	−0.09	−8.9
11	ESO280−SC06	20.34	4.66	26	NGC 3201	0.61	0.74
12	FSR 1735	9.16	0.32	27	NGC 362	3.11	−6.21
13	HP 1	3.59	−0.01	28	NGC 4147	−1.26	18.82
14	IC 1257	5.33	−0.06	29	NGC 4372	2.94	−1
15	IC 1276	8.19	0.3	30	NGC 4590	4.12	6.06

Continue....

#	Name	X	Z		#	Name	X	Z
31	NGC 4833	3.62	−0.92		62	NGC 6293	9.4	1.29
32	NGC 5024	2.83	17.61		63	NGC 6304	5.86	0.55
33	NGC 5053	3.04	17.08		64	NGC 6316	10.33	1.04
34	NGC 5139	3.17	1.34		65	NGC 6325	7.72	1.09
35	NGC 5272	1.48	10		66	NGC 6333	7.73	1.47
36	NGC 5286	7.64	2.15		67	NGC 6341	2.51	4.74
37	NGC 5466	3.35	15.35		68	NGC 6342	8.35	1.44
38	NGC 5634	15.66	19.09		69	NGC 6352	5.27	−0.7
39	NGC 5694	26.43	17.69		70	NGC 6355	9.16	0.87
40	NGC 5824	26.4	12.06		71	NGC 6356	14.76	2.68
41	NGC 5897	10.32	6.3		72	NGC 6362	5.97	−2.29
42	NGC 5904	5.12	5.47		73	NGC 6366	3.19	0.97
43	NGC 5927	6.41	0.65		74	NGC 6380	10.72	−0.65
44	NGC 5946	8.92	0.77		75	NGC 6388	9.52	−1.16
45	NGC 5986	9.32	2.39		76	NGC 6397	2.09	−0.48
46	NGC 6093	9.35	3.33		77	NGC 6401	10.56	0.74
47	NGC 6101	10.97	−4.2		78	NGC 6402	8.38	2.38
48	NGC 6121	2.09	0.61		79	NGC 6426	17.45	5.76
49	NGC 6139	9.56	1.22		80	NGC 6440	8.4	0.56
50	NGC 6144	8.48	2.41		81	NGC 6441	11.48	−1.01
51	NGC 6171	5.88	2.5		82	NGC 6453	11.54	−0.78
52	NGC 6205	2.76	4.65		83	NGC 6496	10.89	−1.96
53	NGC 6218	4.14	2.13		84	NGC 6517	9.94	1.25
54	NGC 6229	6.55	19.73		85	NGC 6522	7.68	−0.53
55	NGC 6235	11.18	2.69		86	NGC 6528	7.88	−0.57
56	NGC 6254	3.91	1.72		87	NGC 6535	5.95	1.23
57	NGC 6256	10.05	0.59		88	NGC 6539	7.24	0.92
58	NGC 6266	6.7	0.87		89	NGC 6540	5.28	−0.31
59	NGC 6273	8.67	1.43		90	NGC 6541	7.23	−1.46
60	NGC 6284	15.06	2.64		91	NGC 6544	2.98	−0.12
61	NGC 6287	9.23	1.8		92	NGC 6553	5.97	−0.32

Continue....

#	Name	X	Z		#	Name	X	Z
93	NGC 6558	7.36	−0.78		119	NGC 7089	5.57	−6.72
94	NGC 6569	10.83	−1.27		120	NGC 7099	4.93	−5.91
95	NGC 6584	12.33	−3.81		121	NGC 7492	7	−23.53
96	NGC 6624	7.82	−1.09		122	Pal 10	3.59	0.28
97	NGC 6626	5.42	−0.53		123	Pal 11	10.97	−3.6
98	NGC 6637	8.66	−1.57		124	Pal 12	11.02	−14.05
99	NGC 6638	9.24	−1.17		125	Pal 13	0.97	−17.63
100	NGC 6642	7.93	−0.91		126	Pal 14	49.7	51.38
101	NGC 6652	9.8	−1.97		127	Pal 15	38.89	18.56
102	NGC 6656	3.13	−0.42		128	Pal 5	16.16	16.65
103	NGC 6681	8.78	−1.95		129	Pal 6	5.79	0.18
104	NGC 6712	6.22	−0.52		130	Pal 8	12.33	−1.51
105	NGC 6715	25.58	−6.45		131	Pyxis	−5.9	4.8
106	NGC 6717	6.8	−1.34		132	Rup 106	10.66	4.29
107	NGC 6723	8.31	−2.59		133	Terzan 1	6.69	0.12
108	NGC 6749	6.37	−0.3		134	Terzan 10	5.78	−0.2
109	NGC 6752	3.31	−1.73		135	Terzan 12	4.75	−0.18
110	NGC 6760	5.96	−0.51		136	Terzan 2	7.48	0.3
111	NGC 6779	4.27	1.36		137	Terzan 3	7.82	1.31
112	NGC 6809	4.9	−2.13		138	Terzan 4	7.18	0.16
113	NGC 6838	2.19	−0.32		139	Terzan 5	6.88	0.2
114	NGC 6864	17.66	−9.08		140	Terzan 6	6.79	−0.26
115	NGC 6934	9.07	−5.05		141	Terzan 7	21.38	−7.82
116	NGC 6981	11.7	−9.18		142	Terzan 8	23.8	−10.93
117	NGC 7006	17.17	−13.69		143	Terzan 9	7.08	−0.25
118	NGC 7078	3.9	−4.77		144	Ton 2	8.08	−0.49

Data obtained from:
http://www.atlasoftheuniverse.com/globular.html

DATASHEET #2

GAS CLOUDS

#	X	Y	#	X	Y
1	1.01	3.41	37	5.63	8.72
2	0.47	3.08	38	5.38	9.65
3	0.05	3.32	39	6.81	11.13
4	−0.71	3.04	40	7.29	10.86
5	−1.13	3.04	41	7.01	10.11
6	−1.47	3.37	42	6.90	10.71
7	−2.25	3.31	43	7.43	9.48
8	−2.98	3.55	44	7.69	8.63
9	−3.69	4.12	45	7.38	8.36
10	−2.93	4.03	46	7.83	8.33
11	−4.02	4.67	47	8.22	7.40
12	−4.47	4.18	48	7.94	6.22
13	−4.89	4.66	49	7.52	6.58
14	−5.18	5.21	50	7.94	6.64
15	−5.62	5.30	51	7.75	5.92
16	−5.32	5.93	52	7.27	5.98
17	−5.96	6.26	53	7.80	5.20
18	−6.07	5.69	54	7.72	3.90
19	−6.61	6.68	55	7.38	3.57
20	−6.81	7.16	56	7.30	2.69
21	−6.36	8.22	57	7.02	2.48
22	−6.44	8.58	58	7.22	1.52
23	−6.64	9.00	59	6.49	2.24
24	−6.81	9.54	60	5.67	0.67
25	−6.87	9.85	61	5.06	−0.50
26	−7.09	7.80	62	4.58	−1.23
27	3.22	4.68	63	4.27	−1.38
28	3.00	4.41	64	3.63	−1.23
29	2.52	4.68	65	2.81	−0.96
30	2.33	4.34	66	1.83	−1.50
31	1.88	4.34	67	−9.84	8.27
32	5.47	5.22	68	−9.48	7.64
33	5.81	5.92	69	−9.31	7.40
34	6.00	6.70	70	−9.28	6.35
35	5.92	7.24	71	−8.74	6.35
36	5.86	7.97	72	−9.00	5.62

Continue....

#	X	Y		#	X	Y
73	−8.80	5.68		111	10.72	4.42
74	−8.24	4.69		112	4.50	−4.27
75	−8.18	4.21		113	3.41	−5.72
76	−8.49	4.06		114	1.93	3.41
77	−8.24	3.57		115	1.51	3.38
78	−7.95	3.76		116	−1.58	3.89
79	−7.39	3.30		117	−0.79	3.68
80	−7.03	2.97		118	−0.54	3.34
81	−6.91	2.19		119	−0.28	2.65
82	−6.63	2.31		120	−0.14	2.23
83	−6.13	2.55		121	−0.96	2.47
84	−5.71	2.58		122	2.41	1.15
85	−6.07	1.56		123	1.71	0.33
86	−5.00	1.11		124	3.68	0.61
87	−4.83	0.38		125	4.38	3.62
88	−3.93	1.08		126	−2.02	0.72
89	−9.10	−0.68		127	−2.46	12.99
90	−7.58	−1.52		128	−2.97	12.32
91	−7.47	−2.12		129	−4.15	12.26
92	−5.47	−2.48		130	−3.86	12.44
93	−3.99	−2.69		131	−4.00	11.81
94	−2.97	−3.71		132	−2.74	9.49
95	0.06	−2.02		133	−2.51	9.19
96	1.07	−2.56		134	−2.91	10.00
97	−12.62	2.84		135	3.59	7.27
98	−11.07	0.80		136	3.90	6.46
99	−10.79	0.80		137	4.18	5.85
100	−10.42	0.47		138	4.26	5.43
101	10.64	3.30		139	−3.81	11.06
102	10.73	2.70		140	−3.61	10.63
103	10.05	2.03		141	−3.78	10.54
104	9.66	1.94		143	−3.64	10.18
105	9.46	1.49		144	−3.58	9.82
106	9.27	0.38		145	−4.37	11.06
107	8.60	−0.80		146	−4.28	9.04
108	8.32	−1.77		147	−4.06	8.52
109	7.42	−2.37		148	−3.04	6.23
110	10.72	5.23		149	−2.11	5.87

Name _____ Id _____

Due Date _____ Lab Instructor _____ Section _____

Worksheet # 1

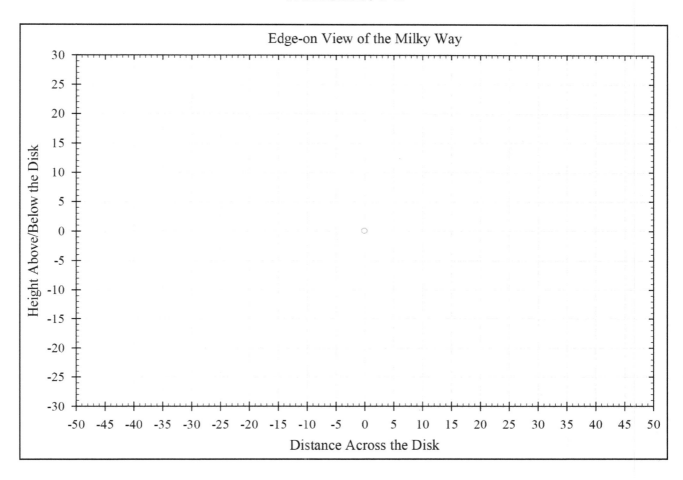

Name _____ Id _____

Due Date _____ Lab Instructor _____ Section _____

Worksheet # 2

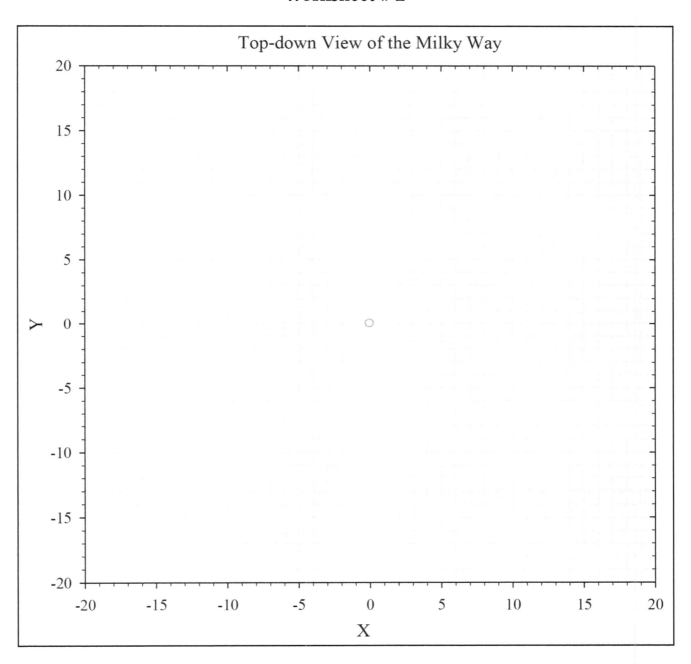

Name _____ Id _____

Due Date _____ Lab Instructor _____ Section _____

Worksheet # 3

Using the positions of globular clusters, determine the mass of the Milky Way galaxy with two methods: Kepler's third law as well as the mass-velocity relationship. Use the equations and conversion factors given in the Equations section. First, where is the center of the Galaxy? The center of the Galaxy should lie at the center of the orbit of the globular clusters. The Sun's position is the (0,0) point on the graph and the positions on Datasheet #1 are measured relative to the Sun's position. Average the height above/below the disk and the distance along the galaxy's plane. Use the Pythagorean Theorem to find the straight-line distance to the Galactic center or use a ruler to estimate the distance from the Sun to the visual center of the clusters' distribution:

Average x-coordinate of the Galactic center in kpc :

Average z-coordinate of the Galactic center in kpc :

Distance from the Sun to the Galactic center in kpc :

Assuming that the Sun's path around the center of the Milky Way is (approximately) circular, what is the circumference of the Sun's orbit (in kpc)?

Circumference of the Sun's orbit in kpc:

Studies of the Sun's orbital motion have shown that the Sun travels around the Milky Way's center at 250 km/sec. How long does it take for the Sun to complete one orbit?

Circumference of the Sun's orbit in kilometers:

Period of the Sun's orbit in seconds:

Period of the Sun's orbit in years:

Kepler's third law states that the period of an orbiting body is determined by the orbital radius and the mass of the central object. For the Milky Way, it can be assumed that this central mass is the combination of the mass of the inner disk and of the galactic nucleus. What is the mass of the Milky Way located inside of the Sun's orbit?

Radius of the Sun's orbit in AU:

Mass of the Milky Way in solar masses:

UNIT 7.6 THE HUBBLE DISTANCE-REDSHIFT RELATION

OBJECTIVE

To learn how Edwin Hubble and the cosmologists of the 20th century used observations of the motions of galaxies to determine the distance to galaxies and the expanding nature of the Universe. You will be able to calculate the Hubble Constant and use recessional velocities to determine extragalactic distances.

INTRODUCTION

In the 1920s, astronomers wrestled with the question of the scale of the Universe and our place in it. An important question to be answered was the nature of the Milky Way, namely, was the entirety of the Universe coinciding with our Galaxy, or was there a larger Universe beyond the Milky Way? Astronomers – with the lower powered telescopes of the 1800's and 1900's – were aware of small smudges in the sky which they called "spiral nebulae," Latin for spiral clouds. However, the question remained whether these smudges in the sky were small clouds orbiting the Milky Way or whether these were whole other galaxies. While philosophers, such as Immanuel Kant in 1755, had theorized that the Universe could in fact be filled with "island Universes," scientific verification of this idea proved elusive.

In 1917, the astronomer Heber Curtis used measurements of stars and globular clusters to find the direction of and distance to the center of the Milky Way galaxy, as well as the first reasonable estimate of the size of the Milky Way's disk. At a staggering 110,000 light-years (34 kpc), it was realized that our Galaxy was truly immense, but it was uncertain as to whether other spiral nebulae were beyond the boundary of the Milky Way. The difficulty in finding their distances is that even the brightest main-sequence stars would be invisible to telescopes of the day at the distance of the edge of the Milky Way and beyond. To determine the distances to spiral nebulae, astronomers would have needed a distance determination technique that stretched all the way out to millions of parsecs.

Astronomer Edwin Hubble settled the debate of the nature of spiral nebulae by using extremely bright, easily detectable Cepheids to determine those smudges in the sky were very distant galaxies. Many of the small spiral nebulae – first noticed by the astronomer Charles Messier and given names like M31, M101, and M77 – turned out to be "island Universes." Using the world's largest telescope, Hubble found M31 (the Andromeda Galaxy) to be located 778,000 pc from the Milky Way (0.778 Mpc). M82 was located 3.62 Mpc, M51was located 7.1 Mpc, and M77 was located 14.4 Mpc away from the Milky Way.

Much more stunning than just the huge distances, Edwin Hubble looked at spectra of the galaxies and noticed that their prominent spectral lines – Calcium-H at 3968.4673 Å and Calcium-K at 3933.6614 Å – were always redshifted. The redshift – which was due to the galaxies' velocity away from the Milky Way – turned out to be highly linear: the further away a galaxy was from our own, the faster it appeared to be moving away. Taking the velocity of the galaxy and dividing it by the distance always yielded the same number, called the Hubble Constant or Hubble Parameter. So, if one galaxy, Galaxy-B, was seen to be moving 5× faster than another galaxy – Galaxy-A – it was because Galaxy-B was also 5× further away. If Galaxy-C was seen moving 2× faster than Galaxy-B, then Galaxy-C was 2× further away than B (and 10× further away than A). The ratio of velocity to distance always held. The value of that ratio is called the Hubble Constant or Parameter.

This was the first evidence of the event we today refer to as the "Big Bang," the event that marked the beginning of the Universe's existence. The first proponent of this foundational event was the Belgian astrophysics and Catholic priest Georges Lemaître. As a side note, the preferred term for this theory at the time was the "primordial atom theory," describing the Universe emerging from a hot, dense, atom-sized state. "Big Bang" was coined by opponents of Lemaître's theory and was supposed to be dismissive by tagging the theory with a silly and childish name. Today, the Big Bang creation of the Universe is widely accepted, considered a "science fact" by most if not all scientists, with the redshift of galaxies – asides from the detection of the Cosmic Microwave Background, corresponding to a black body temperature of about 3 K – being the "smoking gun" evidence to the Universe's continuing expansion and evolution.

EQUATIONS AND CONSTANTS

Equation	Expression	Variables
Redshift	$\Delta\lambda = \lambda_{obs} - \lambda_0$	$\Delta\lambda$: the Doppler shift λ_{obs}: the observed wavelength of a spectral line λ_0: the rest wavelength of a spectral line
Radial Velocity	$v_R = \dfrac{\Delta\lambda \times c}{\lambda_0}$	v_R: the radial velocity in km/sec $\Delta\lambda$: the Doppler shift λ_0: the rest wavelength of a spectral line c: the speed of light in km/sec
Hubble's Law	$H_0 = \dfrac{v}{D}$	H_0: Hubble's Constant v: a galaxy's recessional velocity in km/sec D: the distance to a galaxy in Mpc
Constants and Conversions		
$c = 2.99792458 \times 10^5$ km/sec		
1 Mpc = 1,000 kpc = 1,000,000 pc		
1 Mpc = $3.08567758 \times 10^{19}$ km		
1 year = 3.15576×10^7 sec		

PROCEDURE

Datasheet #1 contains the distances – in Mpc – of six relatively nearby galaxy clusters, whose distances have been determined by the Cepheid period-luminosity relationship.

Worksheet #1 contains spectra from the brightest, most easily spotted galaxy in 6 different galaxy clusters. The prominent spectral line seen in the data is the Calcium-H spectral line. Its rest wavelength is 3968.4673 Å. However, the recessional velocity of each of these galaxy clusters causes the Calcium-H line to be noticeably red-shifted. For each galaxy cluster, record the observed wavelength, calculate the Doppler shift, and calculate the recessional velocity of the galaxy. Record your recessional velocities in the spaces provided on Datasheet #1.

Using your distances (in Mpc) and velocities (in km/sec), plot those data points on Worksheet #2's graph and answer the post-lab questions.

DATASHEET #1

Galaxy Cluster	Cepheid-Determined Distance (in Mpc)
Virgo Cluster	18
Sculptor Cluster	32
Antlia Cluster	41
Perseus-Pisces Cluster	72
Coma Cluster	99
Hercules Cluster	156

Galaxy Cluster	Recessional Velocity (in km/sec)
Virgo Cluster	
Sculptor Cluster	
Antlia Cluster	
Perseus-Pisces Cluster	
Coma Cluster	
Hercules Cluster	

Name _____ Id _____

Due Date _____   Lab Instructor _____ Section _____

Worksheet #1

Virgo Cluster

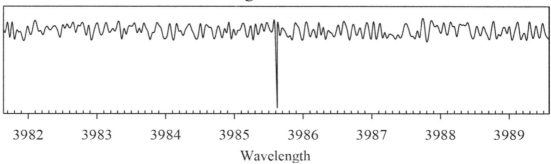

Wavelength

Observed Wavelength (in Å)	Doppler Shift (in Å)	Radial Velocity (in km/sec)

Sculptor Cluster

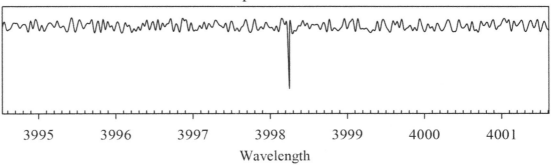

Wavelength

Observed Wavelength (in Å)	Doppler Shift (in Å)	Radial Velocity (in km/sec)

Continue....

Antlia Cluster

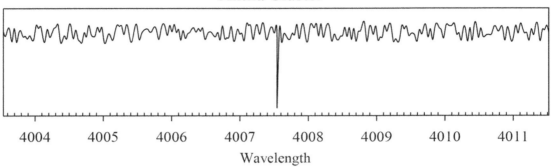

Wavelength

Observed Wavelength (in Å)	Doppler Shift (in Å)	Radial Velocity (in km/sec)

Perseus-Pisces Cluster

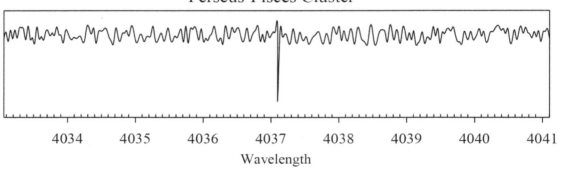

Wavelength

Observed Wavelength (in Å)	Doppler Shift (in Å)	Radial Velocity (in km/sec)

Continue....

Name _____ Id _____

Due Date _____   Lab Instructor _____ Section _____

Worksheet #1, cont'd

Coma Cluster

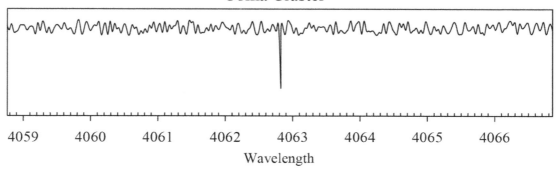

Wavelength

Observed Wavelength (in Å)	Doppler Shift (in Å)	Radial Velocity (in km/sec)

Hercules Cluster

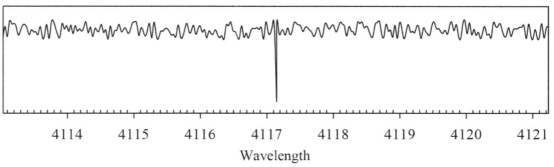

Wavelength

Observed Wavelength (in Å)	Doppler Shift (in Å)	Radial Velocity (in km/sec)

Name _____ Id _____

Due Date _____   Lab Instructor _____   Section _____

Worksheet # 2

Galaxy Distance-Velocity Relationship

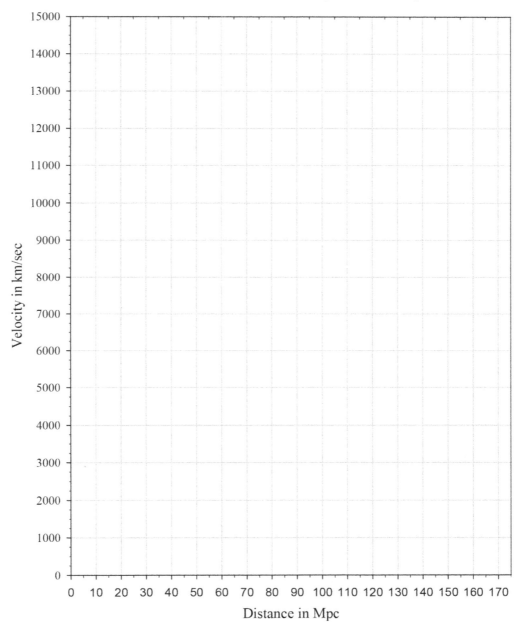

Distance in Mpc

Name _____ Id _____

Due Date _____ Lab Instructor _____ Section _____

Worksheet # 3

Determination of the Hubble Constant

Draw a smooth, straight, best-fit line through the data points, beginning at the point (0,0). Pick a spot on your best fit line (not one of the plotted points) and place an **x** at that spot. Use that point to determine Hubble's constant for the Universe.

Velocity of the point on the best fit line (in km/sec):

Distance of the point on the best fit line (in Mpc):

Your determined value of the Hubble Constant (in $\dfrac{km / sec}{Mpc}$):

A galaxy is seen to be moving away from the Milky Way at a speed of 12,520 km/sec, but doesn't have any visible Cepheids allowing determining its distance. Use your chart and the velocity to determine the galaxy's distance in Mpc.

Distance of the galaxy:

A very, very distant galaxy is seen to be so far away that it appears as little more than a single smudge of dim light. Therefore, astronomers have no hope of seeing Cepheids or bright stars to gauge its distance, but it can be determined, through Doppler shifting, that the galaxy is receding at a velocity of 201,600 km/sec, which is two thirds the speed of light. Use the Hubble's constant equation and your value of H_0 to determine the distance of the galaxy in Mpc.

Distance of the galaxy:

Name _____ Id _____

Due Date _____   Lab Instructor _____   Section _____

Worksheet # 4

Determination of the Age of the Universe

The Big Bang theory states that the Universe had a finite beginning some measureable time in the past, marking the beginning of matter, energy, time, and space, i.e., the Universe as we know it. At one point, everything in the Universe was compacted into a single point (simplified representation), before the Universe began expanding and spreading out the Universe's mass and energy across a vast area. The expansion of the Universe that began at the Big Bang continues today, as astronomers such as Edwin Hubble and Vesto Slipher concluded by watching galaxies recede from the Milky Way, riding along with the still-inflating Universe.

For the nearby Whirlpool Galaxy – M51 – astronomers can use Cepheids to verify its distance of 6.614 Mpc and spectra to measure its recessional velocity of 463 km/sec. This galaxy also fits Hubble's law:

$$H_0 = \frac{v}{D} = \frac{463 \ \frac{km}{sec}}{6.614 \ Mpc} = 70 \ \frac{km}{sec \cdot Mpc}$$

Hubble's constant mixes together two separate distance units: km and Mpc. With 1 Mpc = $3.08567758 \times 10^{19}$ km (that's read as 30.8 quintillion kilometers), what is M51's distance from the Milky Way in kilometers?

km

M51's velocity is 463 km/sec, directed away from the Milky Way, meaning that every second that passes, M51 will be located 463 km further away. Conversely, every second in the past, M51 has been 463 km closer. Using the relationship between time, distance, and velocity:

$$D = v \times T$$

Continue....

How many kilometers closer was M51 when Ronald Reagan became US president (January 20, 1981) than it is today (with 1 year = 3.15576 × 10⁷ sec)?

	km

How many kilometers closer was M51 when the dinosaurs were wiped out 65,000,000 years ago?

	km

How many kilometers closer was M51 when the Earth formed 4,500,000,000 years ago?

	km

How many seconds (or years) did it require for M51 to get to 6.614 Mpc away from the Milky Way (although the underlying physics and interpretation is much more complicated)?

	sec		years

The result from your calculation indicates the age of the Universe, the time when the Milky Way, and M51, and every other galaxy, every photon, and every atom were compressed into one singular point. The Universe we have today is a result of matter receding from the Milky Way – and from any other galaxy as well – for a very long time.

Unit 8: A Little Bit of Astrophysics

UNIT 8.1   THE SOLAR SPECTRUM

OBJECTIVE

To learn to identify spectral lines in the Sun's light. You will be able to determine the make-up of the Sun's topmost visible layer, the photosphere, by determining which elements are preferentially removing specific wavelengths of light.

INTRODUCTION

The white light we detect from the Sun is actually a visual optical illusion; white light is a combination of photons of many different energies and wavelengths exciting optical receptors. The light we see from the Sun is a combination of huge numbers of photons with wavelengths between 3600 Å and 7000 Å (the limits of the human vision's ability to detect); note that 10 Å = 1 nanometer (1 nm). What is not noticed from simple observations of the Sun is that minuscule segments of this light are missing from the spectrum, with drastically lower numbers of photons – such as 3933 Å violet photons – than would be expected from a hot gas of a Sun-like temperature of 5800 K. The source of this missing light is the make-up and nature of the subatomic world.

Atoms are discrete units of matter composed of a core nucleus of protons and neutrons surrounded by halos of electrons (in a simplified interpretation). The electron cloud is not a continuous region but is separated into energy levels, imagined as larger and larger nested spheres centered on the nucleus with empty space between the levels. The further from the nucleus an electron's orbit is located, the higher the energy level it is said to occupy. Like rungs on a ladder, electrons can move to higher energy levels if they absorb energy (which happens when they absorb light or are involved in an energetic collision) or the electrons can fall back to lower energy levels closer to the nucleus (in which case the atom will emit light).

A hot, dense energy source – like the inner layers of the Sun – will produce a continuum (that is, a flow of photons composed of all energies and colors, from infrared through visible to the ultraviolet, and beyond). When passed through a prism and spread out according to photon energy, this continuum – to the human eye – takes on the appearance of a rainbow (with all the photons visible to human vision – low energy red through orange, yellow, green, blue, and high energy violet – present). As the light from those lower layers passes through the visible surface of the Sun – called the solar photosphere – atoms in this relatively cool atmospheric layer will absorb and strip some tiny segments of that light.

In order for an electron to leap to a higher energy level, it requires a specific amount of energy (much like a key – to open a lock – requires specifically shaped teeth that fit the lock). Too much energy (from a short-wavelength photon) or too little energy (from a long-wavelength photon) will fail to be absorbed. But at select wavelengths, the electron will absorb that photon and jump to an available energy level, allowing all other photons to pass freely by. The energy of the missing photons in a spectrum is directly related to the element present to do the absorption; thus an astronomer can be confident that certain wavelengths and colors are missing because those photons have been removed. For example, the wavelengths of 6563 Å, 4861 Å, 4341 Å, and 4102 Å may cause electrons in hydrogen atoms to jump to progressively higher energy levels.

Energy levels are unique to different elements. Like fingerprints, the layout of an element's energy levels is exclusive to that element (and notwithstanding hydrogen, it also depends on the element's ionization stage). The energy needed for an electron to ascend the energy levels of hydrogen will be discernibly different from the energy needed for an electron to ascend the energy levels of helium. Therefore, matching the energy of absorbed light to the known energy levels of specific atoms, astronomers are able to discern details about the star, such as the composition and temperature of its atmosphere. While the Sun is almost entirely hydrogen and helium in composition, those atoms have very few electrons (one for hydrogen, two for helium) and many fewer energy levels available for those electrons. Much more complex atoms – like calcium with 20 electrons, magnesium with 12 electrons, and iron with 26 electrons – will produce many more absorption lines, which will be peppered throughout the Sun's spectrum. While those lines might not be very prominent or strong, they will be numerous, and thus convey significant information.

Measuring and understanding the nature of spectral lines allow astronomers to not only determine the make-up of a star's atmosphere (chemical composition) but also the temperature and density of the atmosphere, the gravitational acceleration at photospheric levels due to the star's mass and radius, the stellar magnetic field strength, and various other properties as well. Spectral line determination and measurement are also critical components of measuring galactic distances as well as finding our place in the Universe. Other applications include measurements of the motions of stars and the presence of orbiting planets.

It is noteworthy, however, that the solar spectrum is both more intricate and more far-reaching. It also contains photons of the ultraviolet (UV), X-ray, infrared (IR), and the Radio regime. These segments of the solar spectrum provide testimony of other regions of the solar atmosphere (e.g., chromosphere, transition region, and corona), which are shaped by relatively complex processes largely associated with the energy deposition provided by magnetic fields. At ground level, sunlight is 44% visible light, 3% UV (with the Sun at its zenith), and the remainder mostly infrared. Thus, Earth's atmosphere blocks about 77% of the Sun's UV, almost entirely in the shorter UV wavelengths, when the Sun is highest in the sky (zenith). Of the ultraviolet radiation reaching the Earth's surface, more than 95% is the longer wavelengths of UV-A, with a small remainder UV-B, but there is essentially no UV-C (the most energetic component of the UV regime). This result is extremely relevant for the possibility of surface life on Earth, and it has played a major role regarding the evolution of the terrestrial biosphere.

Regarding X-ray spectra, observations started in the early 1980s based on NASA's "Einstein Observatory" (HEAO-2), which was the first fully imaging X-ray telescope put into space. It provided formation on highly ionized atoms, especially iron, including variability. Notably, ionization stages between Fe X and Fe XV were observed; here Fe X means that nine electrons have been stripped away from the iron atom. Based on this and subsequent space missions, the spectral regime under investigation was extended to wavelengths of less than a nanometer (nm).

Solar IR radiation containing a large number of molecular spectral lines is – very surprisingly – produced by relatively cool gaseous regions (with temperatures as low as 3000 K) embedded in the much hotter solar chromosphere. The spectra associated with these regions are dominated by numerous carbon monoxide (CO) lines, amounting to nearly 300 lines between 2.29 and 2.50 micrometers (μm). The solar radio flux is produced by the solar corona due to temperatures of up to several millions of Kelvin. The associated wavelengths range from 1 cm to 1 m. The solar radio flux is also shaped by so-called radio bursts, which are caused by remarkable explosive features within the solar corona. The analysis of these bursts (among other information) provides insight into the underlying physical processes, deemed responsible for coronal heating and momentum transfer, and ultimately the generation of the solar wind.

EQUATIONS AND CONSTANTS

Equation	Expression	Variables
Scale Factor	$$Scale\,Factor = \frac{Real}{Ruler}$$	*Scale Factor*: a conversion factor which bridges actual and scale model sizes *Real*: a real-world distance or size, such as nanometers or Angstroms *Ruler*: a distance or size used in a scale model, such as millimeters or centimeters

ILLUSTRATIONS

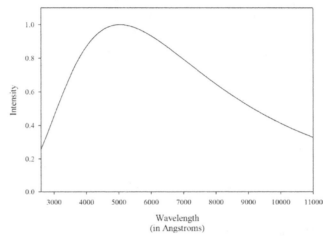

Figure 1. The thermal radiation curve of a perfect black body radiator with a temperature of 5800 Kelvin (like the average temperature of the Sun's photosphere). This graph shows the wavelength (energy) of emitted photons versus the relative number of those photons emitted. The black body curve rises smoothly to a peak at approximately 5000 Angstroms (visible photons seen as blue-green to the human eye) and then drops off sharply at the high energy portion of the spectrum (the invisible ultraviolet photons have wavelengths less than 3600 Å). Human vision is sensitive to photons of wavelengths between 7000 Å to 3800 Å.

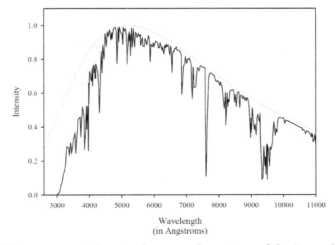

Figure 2. The solid (jagged) line represents the actual measured spectrum of the Sun, with the dotted (smooth) line representing the perfect thermal radiation blackbody curve. The observed curve is created by counting the number and energy of photons which make it to ground-based detectors. The spectrum appears jagged with clusters of sharp dips at specific wavelengths, demonstrating that some types of photons are missing in disproportionately large numbers. Also notice that the number of high energy UV photons making it to Earth-based detectors is much, much lower than the perfect curve owing to the absorption properties of Earth's atmosphere.

PROCEDURE

When recording the spectrum of a star, planet, galaxy, or the Sun, astronomers will introduce known spectral lines into the spectrum for comparison and easy measurement of line positions. In the case of the solar spectrum, the line labeled K is from a calcium atom. The wavelength of those missing photons is known from experiments to have wavelengths of 3933.7 Å. These are the violet visible light photons absorbed when an electron in calcium moves from a lower energy state to a specific excited state. Using Datasheet #1 with prominent, easily spotted solar absorption lines or another mapping of the solar spectrum, you will determine the element which created each of the labeled spectral lines.

For Worksheet #1, before you can measure the position and wavelength of spectral lines in Angstroms (Å), you will need to calculate the scaling of the spectra photograph. The scale factor requires you to compare an actual measurement, in millimeters using a ruler, with the wavelength of the spectral lines, in Angstroms. Measuring from the left end of the line, the right end of the line, or the center of the line doesn't make any difference, so long as you keep your measurements consistent. After calculating a scale factor (from four separate measurements), move on to Worksheet #2 and use this scale factor to determine the identity of the absorption lines.

When measuring spectral lines, use the known spectral line calcium-K (marked as K, with a well measured and verified wavelength of 3933.7 Å) as your starting measurement point. For all measurements that you make, measure from the K line. For the unknown spectral lines (the lines labeled as H, h, g, G, f, e, d, F, c, E, and a), measure their physical distance in millimeters and record those in column 2. These are your ruler measurements used in the scale factor equation.

Your scale factor from Worksheet #1 relates your ruler measurements to actual photon wavelengths. In this case, you will need to convert your ruler separations in column 2 to real separations in Angstroms using the Scale Factor equation. Record that separation in column 3. Remember that your starting point for your measurement was the calcium-K line at 3933.7 Å. Add the separation in column 3 to the calcium-K wavelength to determine the approximate wavelength of the absorption line and record that in column 4.

Finally, using the list of known laboratory spectral lines in Datasheet #2, match your unknown line wavelengths (column 4) to known elemental wavelengths and record your best approximation of the element originating lines H, h, g, etc. Because of measurement error and the limitations of your equipment, the values may not line up perfectly but – if done carefully – will generate values which are very close to the known and accepted wavelengths.

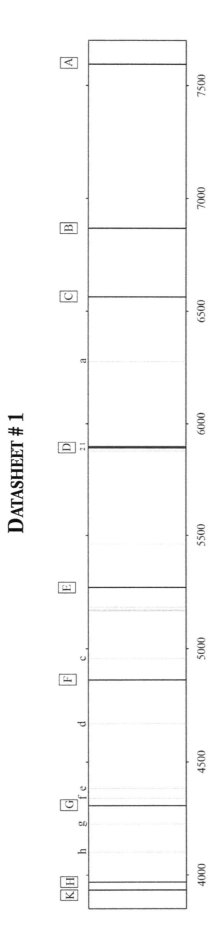

Datasheet # 2

Wavelength	Element	Wavelength	Element
3888.648	He I	4782.991	Si I
3905.523	Si I	4861.342	H
3933.682	Ca II	4920.514	Fe I
3944.016	Al I	4957.613	Fe I
3968.492	Ca II	5084.226	K I
4045.825	Fe I	5088.678	He I
4047.213	K I	5167.327	Mg I
4063.605	Fe I	5172.698	Mg I
4071.749	Fe I	5183.619	Mg I
4101.732	H	5250.216	Fe I
4132.067	Fe I	5269.550	Fe I
4143.878	Fe I	5359.574	K I
4167.277	Mg I	5373.706	Fe I
4226.740	Ca I	5777.756	Mg I
4250.130	Fe I	5782.384	K I
4250.797	Fe I	5875.615	He I
4254.346	Cr I	5889.973	Na I
4260.486	Fe I	5895.940	Na I
4271.774	Fe I	6102.727	Ca I
4307.902	Fe I	6120.270	K II
4340.472	H	6165.590	P II
4383.545	Fe I	6276.606	O_2
4415.135	Fe I	6471.665	Ca I
4455.000	K II	6562.808	H
4471.479	He I	6677.995	Fe I
4528.627	Fe I	6678.151	He I
4554.036	Ba II	6976.523	Si I
4668.140	Fe I	7065.215	He I
4703.003	Mg I	7156.700	O I

Wavelengths generated from data sets chosen from:
http://physics.nist.gov/PhysRefData/ASD/lines_form.html
for the different elements

Name _____ Id _____

Due Date _____ Lab Instructor _____ Section _____

Worksheet # 1

1. Measure: how many millimeters are there between the 5000 and 6000 Å tick-marks?

Real Measure (in Å)	Ruler Measure (in mm)	Scale Factor (in Å/mm)

2. Measure: how many millimeters are there between the 6000 and 6500 Å tick-marks?

Real Measure (in Å)	Ruler Measure (in mm)	Scale Factor (in Å/mm)

3. Lines labeled B and C have wavelengths of 6867.19 and 6562.81 Å, respectively. How many millimeters are there between the B and C lines?

Real Measure (in Å)	Ruler Measure (in mm)	Scale Factor (in Å/mm)

4. Taking an average of your scale factor measurements (in Angstroms/millimeter), how many Angstroms are in each of your measured millimeters in the spectrum?

Average Scale Factor (Å/mm)	

Name _____ Id _____

Due Date _____ Lab Instructor _____ Section _____

Worksheet # 2

Unknown Line Number	Distance from Calcium-K line to unknown line (in mm)	Distance from Calcium-K line to unknown line (in Å)	Wavelength of unknown line (in Å)	Element that produced unknown line
H				
h				
g				
G				
f				
e				
d				
F				
c				
E				
a				

1. Why is the Sun's upper layer – the photosphere – considered "cool" despite its average temperature of 5800 K?

2. The Sun is composed of 99% hydrogen and helium (by numbers), but there are very few helium and hydrogen lines visible in the Sun and many calcium lines, iron lines, and lines from various other metals. Give a well-thought-out explanation as to why the Sun's atmosphere appears dominated by many spectral lines of numerous heavy elements.

3. The so-called element "O_2" on the datasheet is actually molecular oxygen, meaning two oxygen atoms bonded together while also sharing their outer electrons. These microscopic molecular structures can only exist where temperatures are much lower than the 5800 K of the Sun's photosphere. Therefore, the line produced by O_2 absorption could logically not come from O_2 in the Sun. However, astronomers certainly see bands of absorption lines in the solar spectrum caused by molecular oxygen. Explain how these strong molecular oxygen lines could have ended up in spectra of the Sun gathered by Earth-based telescopes.

UNIT 8.2   THE DOPPLER EFFECT

OBJECTIVE

To learn how the position of spectral lines gives astronomers a sense of the movement and motions of objects in space. By calculating how the wavelengths (or frequencies) of photons have shifted from their at-rest values, it will become apparent how radial motions may compress or expand the wavelength of photons.

INTRODUCTION

The Doppler effect was first proposed by Christian Doppler in 1842, an Austrian physicist with respect to sound (acoustic waves). It is commonly heard when a vehicle sounding a siren or horn approaches, passes, and recedes from an observer. Compared to the emitted frequency, the received frequency is higher during the approach, identical at the instant of passing by, and lower during the recession. Regarding the wavelength, the behavior is quite the opposite: the received wavelength is lesser during the approach, identical at the instant of passing by, and increased during the recession.

However, in astronomy and astrophysics, the Doppler effect is mostly discussed and utilized with respect to visible light as well as the other segments of the electromagnetic spectrum. Therefore, in the following, this will be our main focus. Specifically, we will investigate and explore the Doppler effect with respect to distinct spectral lines. The motion of objects becomes obvious by the effect that motion has on the star's spectrum. Spectral lines not only tell astronomers about the composition of a light emitting object, but also the motion of that object.

The word velocity does not simply refer to how fast an object is moving, but also accounts for the direction in which that speed is oriented. Velocities are given in units of length per time (for example, miles per hour or kilometers per second). The velocity describes how a distance between objects changes over time. A 90 kilometer per second velocity translates as a distance either growing or shrinking by 90 kilometers per second. Velocities can be either negative or positive, depending on the orientation of the moving object. A baseball thrown has a velocity that is dependent on who is observing the ball. The observer throwing the baseball may see the ball traveling at +100 mph, as the ball is traveling away from her/him and its distance is increasing (+) as it travels. The person being thrown at will observe the ball traveling at –100 mph. The negative sign is indicative of an object that is decreasing its distance from the observer. Depending on where an observer is standing, she/he will register a different velocity for the baseball in motion. An observer standing aside watching the ball fly directly across her/his line of sight will calculate a velocity of 0 m/sec (since the baseball is neither getting closer to that person nor further away from her/him, simply passing in a straight line through his field of view).

Velocities can have a measureable effect on the way we perceive moving objects. The motion of an object can distort the wavelength of absorbed or emitted photons in much the same way as the motion of a car will distort the sound of its horn to a listener standing on the street. The sound of a car horn will be distorted by the car's approach (where the frequency or pitch of the horn will increase) or by the car's recession (where the frequency or pitch of the horn will decrease) for a listener. Motion *toward* an observer will *compress* sound waves and lead to *shorter* wavelength sound waves (a higher "pitch" sound). Motion *away* from an observer will stretch out sound waves and lead to *longer* wavelength sound (a lower "pitch" sound). Photons can undergo the same expansion and compression; therefore, giving vital information about the motion of stars, gas clouds, and entire galaxies. The speed of light is a fixed quantity and the speed at which photons travel cannot be altered by motion. For example, a star moving toward the Earth at 50% the speed of light will not emit photons which are traveling 50% faster than the speed of light. The same is true of soundwaves, which do not travel faster or slower because of the motion of the emitting source of the sound. However, the motion of the object emitting sound or light waves is still obvious and measurable because the motion does have an effect on the light or sound waves.

Doppler shifting describes the properties of waves undergoing motion. Waves can be compressed or expanded, depending on whether an object is moving toward or away from an observer. If a star is moving toward an observer, that stellar light will still travel at its constant speed (the speed of light) but its wavelength will be compressed (the energy of the moving object will be expressed by a compression of the light wave peaks). Every emitted wavelength will become "blue-shifted" (with the photons moving to shorter wavelengths, with blue considered "short" wavelength visible light). If a star is moving away from an observer, the velocity of the star will elongate the photon wavelengths; longer wavelengths mean "redder" photons and hence "red-shifted light." So the well-known and laboratory-measured spectral lines of a star will occur at longer wavelength or shorter wavelength, respectively. Since astronomers know the wavelength of certain spectral lines to very high precision, such as the hydrogen alpha line, at 6563 Å. By comparing these rest, unshifted wavelengths to Doppler shifted wavelengths, astronomers can make measurements of the actual motion an object is undergoing, even though this approach is restricted to radial motions (see below).

The Doppler shift is critical to astronomy given the huge distances and small apparent sizes of objects in the Universe. For example, stars orbit the center of galaxies at high velocities but – given the size of a star's orbit – it may take hundreds of millions of years to complete one orbit. Therefore, over the course of a human lifetime, the largest telescope on Earth would not see a star in a galaxy move even the smallest amount. However, the star's velocity can still be calculated by measuring the shift in the stellar light from its rest wavelength pertaining to known nearby stars. Astronomers can use the presence of multiple sets of spectral lines to determine that a single blob of light may actually represent two orbiting stars. The rhythmic shift in spectral lines of a star may perhaps also be the telltale sign of an unseen component tugging on the star, such as the minor tug of an extra-solar planet or the powerful pull of an orbiting black hole companion.

EQUATIONS AND CONSTANTS

Equation	Expression	Variables
Doppler Shift	$\Delta\lambda = \lambda_{obs} - \lambda_0$	$\Delta\lambda$: the Doppler shift λ_{obs}: the observed wavelength of a spectral line λ_0: the laboratory, rest wavelength λ_0: the laboratory, rest wavelength
Radial Velocity	$v_R = \dfrac{\Delta\lambda \times c}{\lambda_0}$	v_R: the radial velocity $\Delta\lambda$: the Doppler shift λ_0: the laboratory, rest wavelength c: the speed of light
Constants and Conversions		
$c = 299{,}792.458 \dfrac{km}{s} \approx 300{,}000 \dfrac{km}{s}$		

ILLUSTRATIONS

Figure 1. Each arrow represents the speed and direction of three objects moving toward the Earth. All three arrows are the same length, representing that all three objects are traveling at the same speed (for example, 100 km/sec). However, Doppler shifting only accounts for motion direct toward or away from an observer's line of sight and can only measure the speed of an object's direct approach. Because those three are approaching at different angles, they are closing the distance to Earth at different rates and thus would show different radial velocities. The bottom object's velocity arrow points directly at Earth, so its full 100 km/sec speed is directed toward closing the gap with Earth (the dashed line demonstrates the distance covered in one second). The middle object's path is notably deviated so the distance it covers is split along two paths: some of the path directed toward the Earth (the dashed horizontal line; *radial velocity component*), and some of it is directed across the Earth's line of sight (vertically; *transverse velocity component*). The top object approaches Earth at a very steep angle, so only a small portion of its velocity carries it toward the Earth, thus making its approach velocity comparatively low.

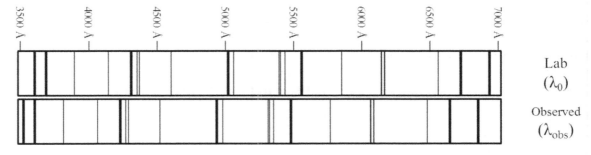

Figure 2. Two absorption spectra aligned with one another. The top absorption spectrum was created from an unmoving laboratory source, showing elemental spectral lines at their rest wavelengths. The bottom spectrum is produced by the same elements but all of the lines are shifted due to the motion of the light generating object. This is a blue-shift, visible in that the spectral lines have moved from longer wavelengths at λ_0 to shorter wavelengths (λ_{obs}). For example, the very first dark spectral line in the top spectrum has a wavelength of ~3600 Å; in the observed spectrum, that line has shifted to 3500 Å, a Doppler shift of ~100 Å.

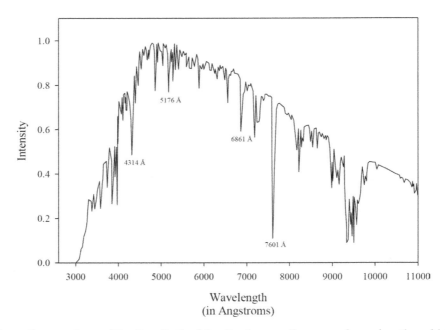

Figure 3. The absorption spectrum of the Sun. Each of the dips have well-measured wavelengths, which are caused by absorption by electrons of elements in the Sun's atmosphere (and would appear as dark bands of "missing" colors in the rainbow of light coming from the solar surface). These wavelengths (4314 Å, 5176 Å, 6861 Å, and 7601 Å) can be considered rest wavelengths and could be reproduced in a laboratory. Any shift in these lines (to longer or shorter wavelengths) would be an indication of radial motion between the Sun and the Earth. For example, if the Sun was moving away from the Earth at +50 km/sec, these lines would shift to longer wavelengths (4314.7 Å, 5176.9 Å, 6862.1 Å, and 7602.3 Å).

PROCEDURE

Below, find seven separate spectra obtained from telescopic observations of seven separate stars. A short description will inform you of the spectral line the astronomer was studying, including its origin and, importantly, the spectral line's rest wavelength. For each of the following spectra, calculate the radial velocity (relative to the observer) of the light-producing object and determine from the list below what was happening to that star to produce that line and the observed shift:

A. **Earth-Sized Planet:** A small, Earth-sized planet's gravity will tug on its parent star and cause the star to wobble back and forth in orbit at a low speed, around 0.05 km/sec.

B. **Jupiter-Sized Planet:** A massive, Jupiter-sized planet's gravity will tug on its parent star and cause the star to wobble back and forth in orbit at a moderate speed, around 0.6 km/sec.

C. **Pulsating:** A pulsating star like a Cepheid will expand and contract its entire surface regularly at between 10 to 100 km/sec.

D. **Flaring:** A solar flare – the powerful eruption of gas from the surface of a star as, e.g., the Sun – will travel upwards into space at a speed of, for example, 500 km/sec.

E. **Binary Member:** Stars orbiting another star in a close binary will travel anywhere from 50 to 300 km/sec through orbit; often, spectral lines of both stars will be visible and shifted.

F. **Black Hole Companion:** A star orbiting close to a black hole will be whipped through orbit at incredible speed, in excess of 2500 km/sec.

G. **Supernova Explosion:** An exploding star undergoing a supernova will blast its outer layers into space at gigantic speeds, such as 25000 km/sec or more.

Name _____ Id _____

Due Date _____ Lab Instructor _____ Section _____

Worksheet # 1

Spectrum # 1

This spectrum is centered on the 5875.6148 Å spectral line found in He I, which is prominent in the photospheres of stars roughly 1000 K hotter than the Sun. Calculate the velocity of the gas producing this line and – from the choices in the datasheet – comment on the nature of this object.

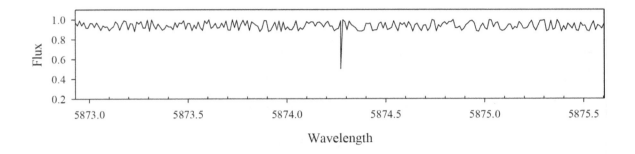

Observed Wavelength (Angstroms)	Shift (Angstroms)	Radial Velocity (km/sec)	Phenomenon

Spectrum # 2

The molecule TiO_2 (titanium dioxide) is a common compound found on Earth (it is the shiny white flecks you find in oil paint, the glaze on pottery, and a major component of sunscreen, since it reflects sunlight from your skin). It is also a complex molecule found in cool, thin atmospheres of low-temperature stars. It has a strong spectral line at the rest wavelength 6680.81 Å.

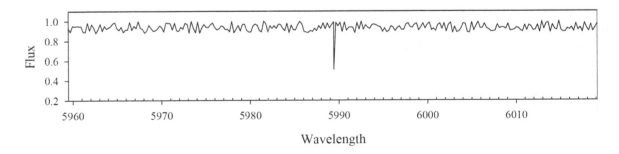

Observed Wavelength (Angstroms)	Shift (Angstroms)	Radial Velocity (km/sec)	Phenomenon

Continue....

Spectrum # 3

This spectrum contains the 3968.4673 Å spectral line associated with the calcium-H line found prominently in the atmospheres of stars with the same surface temperature as the Sun.

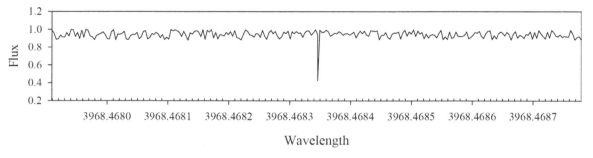

Observed Wavelength (Angstroms)	Shift (Angstroms)	Radial Velocity (km/sec)	Phenomenon

Spectrum # 4

This spectral line is produced by the extremely hot gas normally found in the swirling, energetic corona of a star as the Sun. This line is produced by Fe XXI (an iron atom missing 20 of its 26 electrons) and is normally at a wavelength of 1354.08 Å, in the "Extreme-Ultraviolet" (EUV) region of the spectrum.

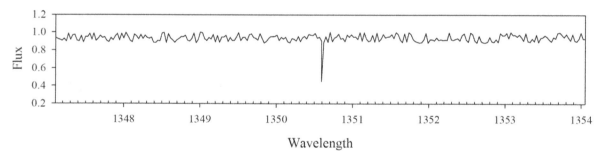

Observed Wavelength (Angstroms)	Shift (Angstroms)	Radial Velocity (km/sec)	Phenomenon

Continue....

Worksheet # 1, cont'd

Spectrum # 5

This spectral line is related to the cyan-colored line of hydrogen found at a rest wavelength of 4340.462 Å. This line is prominent in the atmospheres of very hot (and very young) stars.

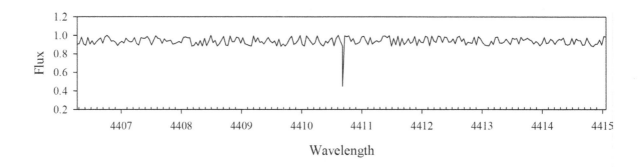

Observed Wavelength (Angstroms)	Shift (Angstroms)	Radial Velocity (km/sec)	Phenomenon

Spectrum # 6

This spectral line belongs to hydrogen, normally visible in the red regime with a laboratory wavelength of 6562.8518 Å. This spectral line can, for example, be traced to the hot hydrogen gas in the atmosphere of an 8000 K star.

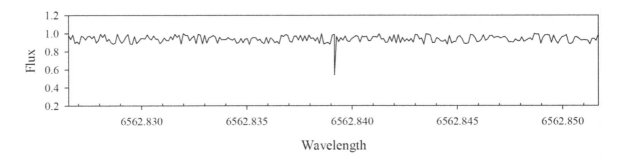

Observed Wavelength (Angstroms)	Shift (Angstroms)	Radial Velocity (km/sec)	Phenomenon

Continue....

Spectrum # 7

This spectrum shows two spectral lines, the right-hand line traced back to Mg II, found prominently in 4500 K gas, with a rest wavelength of 4481.126 Å. The left-hand line is related to helium gas with a temperature of 11,000 K, with a rest wavelength of 4471.479 Å.

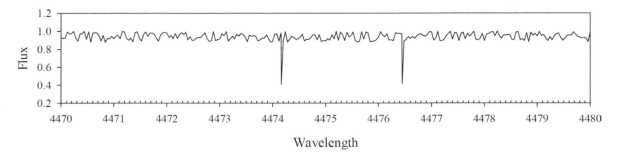

Observed Wavelength (Angstroms)		Shift (Angstroms)	Radial Velocity (km/sec)	Phenomenon
He				
Mg				

Name _____ Id _____

Due Date _____ Lab Instructor _____ Section _____

Worksheet # 2

1. What types of waves exhibit the Doppler effect?

    a) Light waves
    b) Water waves
    c) Sound waves
    d) All of the above
    e) None of the above

2. What is the likely reason that the Doppler effect has first been proposed for sound waves rather than light waves? (Okay, granted, this also answers some of question 1.)

3. In regard to electromagnetic waves, what can be measured by the Doppler effect?

    a) Radial velocities
    b) Transverse velocities
    c) Total (combined) velocities

4. In regard to electromagnetic waves, which part(s) of the electromagnetic spectrum are subjected to the Doppler effect? Name three examples.

5. Residing on the assumption that the Doppler effect is found both for sound waves and electromagnetic waves (in vacuum) – what (if any) is the role of the medium? In which cases (if any) is the underlying equation affected depending on whether the observer moves or the source moves?

UNIT 8.3 THE COLORS AND LUMINOSITIES OF STARS

OBJECTIVE

To learn how astronomers can use simple observations of stars to determine temperatures and how those temperatures can be used to determine the stellar light output.

INTRODUCTION

Any object made of solid, liquid, or dense gas with a well-defined temperature will radiate energy into space in the form of countable energy-carrying photons. These objects are called thermal radiators (named for the heat energy they liberate); they emit energy in a very predictable manner. From the hottest star to the coldest cloud of intergalactic dust, all of these objects shed heat energy by radiating light into space in the form of energetic photons, carrying away their energy as they try to reach a temperature-equilibrium with their surroundings. The energy of individual photons, the number of photons emitted, and the total energy shed by an object depends on the temperature of the object. Objects which are radiating away energy do not do so in a random, irregular, or variable manner. Instead, emitters are bound to laws of physics which govern *how much* and *what kind* of energy can be emitted. As an object's temperature increases, its light output increases (the number of photons) and the energy of the photons emitted intensifies. This type of light output is called a continuous blackbody spectrum or a thermal radiation curve. (Several thermal radiation curves are shown in Figure 2.)

Energy emission is calculated as the number of photons emitted per second per square meter of surface. For example, a one meter by one meter square of 5800 K gas (10,000° F, like the visible photosphere layer of the Sun) produces a gargantuan amount of energy in the form of photons. Each second, that one-meter by one-meter gas would emit 22 trillion 6500 Å photons (the energy which stimulates the red color receptors in the eye). That same gas produces 23.7 trillion photons at 6000 Å (registered by the eye as the color orange). 24.9 trillion yellow (5700 Å) photons are produced each second, in addition to 25.5 trillion green photons (the same color as a green laser pointer, with photons of wavelength 5300 Å). Technically, the Sun produces more green photons across its surface than it does yellow photons. The Sun's peak photon output falls at 5000 Å (a color on its own which your eye would distinguish as turquoise or cyan in color), where each square meter of the solar surface produces 26 trillion of these photons. The photon output then drops sharply at high energies. For example only 10 trillion 2900 Å photons (damaging, sunburn-causing ultraviolet radiation referred to as UVB) are emitted per second per square meter. Moreover, less than 1 billion highly damaging 1100 Å UVC photons (the type of photon used in hospitals to destroy germs and contagions) are produced per square meter. The Sun's black body curve is illustrated in Figure 1.

Generally, constructing a full thermal radiation curve would require the counting of the number of each photon according to its wavelength and – finding the peak of the curve – would yield the stellar surface temperature. This is a difficult proposition, however, especially with the technology available early in the 20th century. Rather than trying to construct a thermal radiation curve of a star by counting the number of incoming photons of each wavelength, astronomers are able to take a shortcut and still determine a star's surface temperature. Telescopes may be equipped with "filters," specially coated glass that allows only specific wavelengths of photons to pass through and be recorded. The "V" of "visual band" filter is so named because it allows through the photons which human vision is most sensitive to (namely, teal, green, and yellow photons). Photons with wavelengths between 5000 Å and 6000 Å may pass through this filter and be recorded. All other light is absorbed, reflected, or scattered away. Photographs made with the V-filter most closely resemble what the human eye would see through a telescope, so the brightness a star appears to have to the naked eye is usually called its "visual magnitude." The "B" filter or "blue band" filter allows through only high energy visible light, the colors humans see as violet, indigo, and blue. The "R" filter or "red band" filter blocks all wavelengths except those the human eye would recognize as orange or red. By measuring the blue magnitude (with a B-filter) and the visual magnitude (the apparent magnitude, through the V-filter) astronomers could use the difference in light output to determine the star's temperature. This value was called the *color index*, sometimes written as (B–V), symbolizing that the color index is the B-magnitude value minus the V-magnitude value. Table 1 lists color indices and the corresponding temperature (in Kelvin) along with the general appearance in star color in a photograph (though the color of a star through a telescope is usually much more subtle).

Before the advent of telescopes and recording devices (such as film or photomultiplier tubes), measuring the brightness of stars was a difficult and somewhat arbitrary process. The ancient Greek mathematician Hipparcos constructed a ranking scheme for starlight called the magnitude system. The brightest stars in the sky were lumped together as "1st magnitude," despite the fact that some first magnitude stars – like Sirius, in Canis Major – were clearly brighter than other first magnitude stars, like Rigel, in Orion. The next grouping of bright stars were classed by Hipparcos as "2nd magnitude" followed by the much dimmer "3rd magnitude" all the way down to the limit of what the naked eye could detect, the 6th magnitude. Therefore, the magnitude system is backwards, meaning that a higher magnitude number translates to a lower light output. Later, instruments that allowed for the much more precise measurement of the amount of incoming light allowed for the refinement of the magnitude system. Rather than Sirius, Altair, and Vega all being referred to as "1st magnitude" stars without taking into account the intrinsic brightness difference of these three stars, astronomers could give magnitudes meaningful values.

A difference in 1 magnitude translates to a difference of 2.512×in light output. For example, if an observer received 1,000,000 photons from a 3.0 magnitude star, the same observer would receive 2.512×as many photons (2,512,000 photons) from a 2.0 magnitude star. In turn, a 1.0 magnitude star is 2.512 times brighter than a 2.0 magnitude star and the observer would receive 6,310,000 photons from the 1.0 magnitude star. A difference on 5 magnitudes is equal to a 100×difference in light output. The nearby star Vega serves as the zero-point for the magnitude system, with its apparent magnitude set to 0.0 and the brightness of all other stars measured relative to Vega. Therefore, the apparent magnitude of a star is a measure of how much of that star's light is reaching an observer; this value is strongly influenced not only by the star's actual light output but also by its distance. The further away a star is from the Earth, the less of its generated light will reach our planet.

Finally, a star's actual light output – called its luminosity – is measured by a value called the absolute magnitude. The absolute magnitude is a measure of how bright a star would appear if it was located 10 pc from the Earth. The absolute magnitude is normally a very difficult value to determine. But for stars which are close to the Earth (i.e., close enough to have their distance determined), astronomers can calculate the absolute magnitude of those stars. By plotting a star's temperature versus its luminosity, astronomers were able to identify a tight correlation between luminosity and temperature. In broad general terms, astronomers found that the higher the star's temperature, the more luminous the star – a statement which was however later recanted due to the discovery of evolved stars, such as giants and supergiants.

EQUATIONS AND CONSTANTS

Equation	Expression	Variables
Color	$(B{-}V) = m_B - m_V$	$(B{-}V)$: the color index m_B: the blue-filter apparent magnitude m_V: the V-filter apparent magnitude
Brightness	$B = 2.512^{m_{dim} - m_{bright}}$	B: the ratio of brightness m_{dim}: the apparent magnitude of the dimmer of two stars m_{bright}: the apparent magnitude of the brighter of two stars
Temperature-Black Body Peak (Wien's Law)	$\lambda = \dfrac{b}{T}$	λ: the wavelength of the black body curve peak in Angstroms b: a constant $= 2.8978 \times 10^7$ T: the temperature in Kelvin

Table 1

Color Appearance	Temperature Range (K)	Color Index
Bluish-White	> 9,800	< 0.00
White	7,400 – 9,800	0.28 to 0.00
Yellowish White	6,100 – 7,400	0.55 to 0.29
Yellow	5,200 – 6,100	0.80 to 0.56
Orange	3,900 – 5,200	1.37 to 0.81
Red	< 3,900	> 1.37

ILLUSTRATIONS

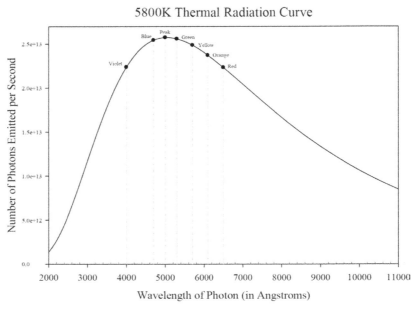

Figure 1. The thermal radiation curve (blackbody curve) of a 1 m² chunk of gas with the same temperature as the visible photosphere of the Sun ("solar surface"). Photons with wavelengths from 2000 Å (UV radiation) through 11000 Å (infrared radiation) are shown. The peak radiation emission by this temperature gas is 5000 Å. The blackbody curve exhibits this general shape regardless of the temperature: a gradual rise through the low energy portion of the spectrum, a peak, and a sharp fall-off at the high energy portion of the spectrum.

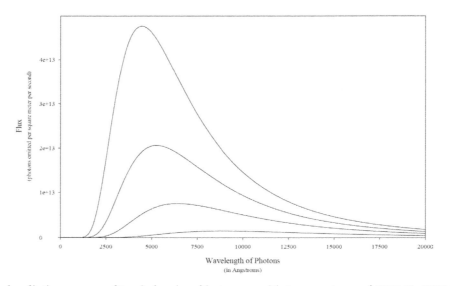

Figure 2. Thermal radiation curves of 1 m² chunks of hot gases with temperatures of 3500 K, 4500 K, 5500 K, and 6500 K, respectively (from lowest to highest). The photon wavelengths, in Angstroms, are plotted along the x-axis. The y-axis displays the number of photons emitted. The 6500 K gas radiates the greatest number of photons at every wavelength, with its peak output at 4450 Å (an energy that our eyes would interpret as dark-blue/indigo in color). According to the y-axis, the 6500 K gas is emitting 47 trillion of photons per second (4.7×10^{13} photons). The 5500 K gas's light output is far lower in comparison. The 5500 K gas's peak wavelength falls at 5300 Å (registered by the eye as green light) and the gas produces 20 trillion of photons per second. Notice that each temperature gas has its own unique energy radiation peak.

PROCEDURE

For Worksheet #1, you will need to use a planetarium, night-sky simulation software, or an astronomical catalogue to gather information about constellations and their brightest stars. Sketch the constellation and mark which star is the main star (this would be the brightest in most cases or one of the brightest stars in that region of the sky). Record the name of the constellation as well as the name of the bright star. In addition, gather the star's apparent magnitude, the star's color (from the list of apparent colors given in Table 1), and the star's luminosity class.

For each of the following, record the name of the constellation shown on the planetarium dome. Roughly sketch the constellation, record the name of the constellation, the name of its main star, the star's magnitude, the star's color, and the star's luminosity class.

Worksheet #2 explores the correlation between the peak radiation wavelength and the gas's temperature. Consider that a radiating body will have a preferential energy regarding the emitted photons, with the wavelength of those photons dependent on the gas's temperature. Referring to the four black body curves presented in Datasheet #2, mark the peak of the black body and determine the wavelength of those photons. Using Wien's Law and the wavelength of those preferentially emitted photons, calculate the gas temperature. Using Table 1, determine what color the gas would appear.

Worksheet #3 deals with the correlation between a star's temperature and its luminosity (absolute magnitude). Using the list of very nearby stars (stars whose temperatures (B–V) and distances can be measured and whose absolute magnitudes can be calculated), graph the color index and the absolute magnitude for those stars. If a correlation between color and luminosity exists, draw a best-fit line through those points.

DATASHEET # 1

#	Star Name	Apparent Magnitude	Absolute Magnitude	(B–V)
1	ξ Bootis	6.80	7.67	1.17
2	τ Ceti	3.45	5.64	0.73
3	π³ Orini	3.15	3.62	0.48
4	ε Indi	4.65	6.86	1.04
5	α PsA	1.15	1.72	0.13
6	82 Eridani	4.25	5.34	0.70
7	ζ Tucanae	4.20	4.53	0.57
8	GX Andromedae	8.05	10.28	1.55
9	Gliese 105 A	5.75	6.47	0.92
10	36 Ophiuchi	5.00	6.11	0.85
11	α¹ Centauri	0.10	4.45	0.60
12	χ Draconis	3.55	4.02	0.48
13	α Canis Majoris	−1.45	1.44	0.00
14	α Aquali	0.75	2.20	0.22
15	o² Eridani	4.40	5.91	0.82
16	α² Centaurus	1.20	5.55	0.82
17	α Lyrae	0.00	0.57	0.00
18	61 Cygni A	6.00	8.28	1.07
19	61 Cygni B	5.20	7.49	1.30

Stars chosen at random from:

https://en.wikipedia.org/wiki/List_of_nearest_stars_and_brown_dwarfs

Data obtained from:

http://wwwadd.zah.uni-heidelberg.de/datenbanken/aricns/gliese.htm
http://simbad.u-strasbg.fr/simbad/

DATASHEET # 2

Blackbody Number 1

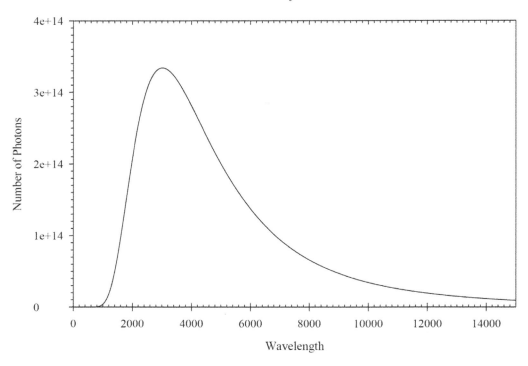

Blackbody Number 2

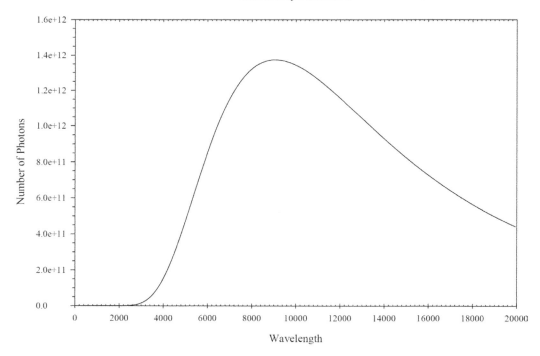

Continue....

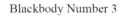

Blackbody Number 3

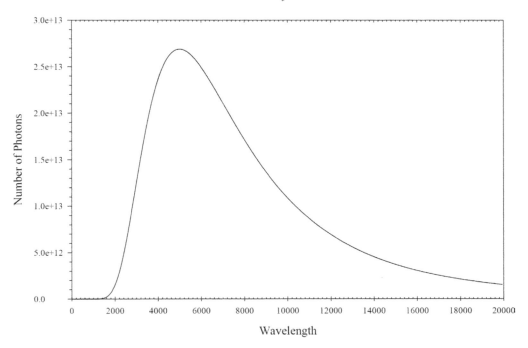

Blackbody Number 4

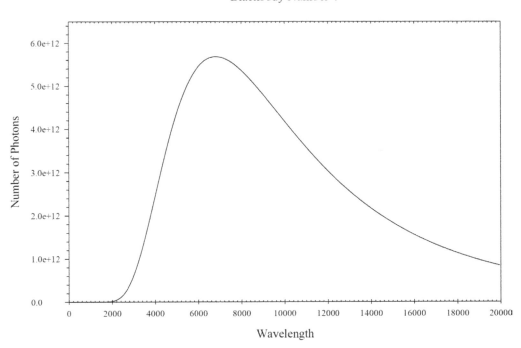

Name _____ Id _____

Due Date _____ Lab Instructor _____ Section _____

Worksheet # 1

Sketch #1		
	Constellation:	
	Main Star:	
	Main Star Magnitude:	
	Main Star Color:	
	Main Star Size Class:	

Sketch #2		
	Constellation:	
	Main Star:	
	Main Star Magnitude:	
	Main Star Color:	
	Main Star Size Class:	

Sketch #3		
	Constellation:	
	Main Star:	
	Main Star Magnitude:	
	Main Star Color:	
	Main Star Size Class:	

Sketch #4		
	Constellation:	
	Main Star:	
	Main Star Magnitude:	
	Main Star Color:	
	Main Star Size Class:	

Continue....

Sketch #5

Constellation:	
Main Star:	
Main Star Magnitude:	
Main Star Color:	
Main Star Size Class:	

Sketch #6

Constellation:	
Main Star:	
Main Star Magnitude:	
Main Star Color:	
Main Star Size Class:	

Sketch #7

Constellation:	
Main Star:	
Main Star Magnitude:	
Main Star Color:	
Main Star Size Class:	

Sketch #8

Constellation:	
Main Star:	
Main Star Magnitude:	
Main Star Color:	
Main Star Size Class:	

Name _____ Id _____

Due Date _____ Lab Instructor _____ Section _____

Worksheet # 2

Black Body Curve #1

Peak Wavelength (in Å)	Surface Temperature (in K)	Color

Black Body Curve #2

Peak Wavelength (in Å)	Surface Temperature (in K)	Color

Black Body Curve #3

Peak Wavelength (in Å)	Surface Temperature (in K)	Color

Black Body Curve #4

Peak Wavelength (in Å)	Surface Temperature (in K)	Color

Name _____ Id _____

Due Date _____ Lab Instructor _____ Section _____

Worksheet # 3

Color vs. Absolute Magnitude

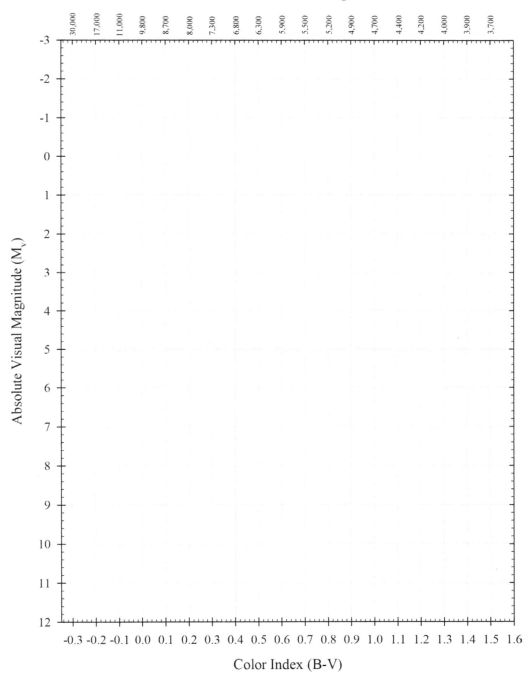

Name _____ Id _____

Due Date _____ Lab Instructor _____ Section _____

Worksheet # 4

These post lab questions refer to the 8 stars from Worksheet #1.

1. Which star appears brightest? Which star appears dimmest? Using the brightness equation, how much more light do we receive from the brightest star as compared to the dimmest?

2. Which star has the highest surface temperature?

3. Which star has the lowest surface temperature?

4. The Sun has a (B–V) color index of 0.62. Which star or stars (if any) likely also have a color index close to 0.62?

5. Vega has a (B–V) of 0.00. Which star or stars (if any) have a color index with a negative value?

Name _____ Id _____

Due Date _____ Lab Instructor _____ Section _____

Worksheet # 5

These post lab questions refer to the stars from Datasheet #1 and the graph of color index versus luminosity that you graphed on Worksheet #3.

1. What is the name of the coolest nearby star and what tells you this?

2. The hottest nearby star has the proper name Vega, but on the list of stars on Datasheet #1, it goes by a different, secondary name. What is that other name of Vega?

3. What is the name of the most *luminous* nearby star and what led to that conclusion?

4. What particular physical characteristics of stars do (B–V) and M_V measure?

5. Do you see any trend in your data? Explain coherently that trend.

6. The m_V of the Sun is measured to be –26.7 and its m_B is determined as –26.05. Therefore, what is the color index of the Sun?

7. Using your graph and the color index you just calculated for the Sun, what would be the absolute magnitude of the Sun?

8. Imagine an astronomer found a star so far away that she or he couldn't measure its distance and thus had no reasonable way of calculating its absolute magnitude. However, m_B and m_V could be determined as 25.9 and 24.5, respectively. What is the star's absolute magnitude?

UNIT 8.4   SPECTROSCOPIC BINARIES

OBJECTIVE
To learn how astronomers can use spectral lines to determine the presence and motion of orbiting binary stars

INTRODUCTION
Nearly half of the stars in the night sky are actually binary stars, meaning two stars which orbit one another. Our Sun is a single star, orbited by a family of eight planets, several dwarf planets, asteroids, comets and other leftover material from the Solar System formation process. Many familiar, bright, and easy-to-spot stars in the sky are binary stars. The brightest star in the night sky – Sirius, located in the constellation Canis Major – is a binary (a double star system). The closest star to the Sun – Alpha Centuri in Centaurus – is actually a triple star system. The North Star – Polaris in Ursa Minor – is also a hierarchic triple star system, with one large bright star orbited by two much smaller and dimmer stars. Furthermore, the stars Mizar and Alcor in Ursa Major are actually a sextuple star system, with two pairs of binary stars orbiting one another (Mizar Aa and Mizar Ab) while being orbited by a second pair of binary stars (Mizar Ba and Mizar Bb). Those four, in turn, orbit around Alcor A and Alcor B.

Some of these stars are far enough from one another that telescopes are able to resolve each of the stars into separate points of light. The star Albireo at the tip of the constellation Cygnus is a favorite target for amateur stargazers because of the striking color difference that can be seen through a small telescope, with the bright orange Albireo A and the blue Albireo B orbiting one another over a nearly 100,000 year period.

For a telescope to resolve a binary into individual stars, though, the stars must either be viewed with a very large telescope or the stars must be very far from one another. For example, orange and blue Albireo A and B are only seen as individual stars because they are about 4100 AU apart (i.e., more than 100 times the Sun-Pluto distance). If the two binary components aren't many thousands of billions of kilometers from one another, then even the largest telescope will show the binary system as being nothing more than a single blob of combined light. Albireo A is, itself, a binary star, with the gigantic bright orange star orbited by a much smaller, closer blue star; these stars are approximately 27 AU apart (thus, strictly speaking, Albireo constitutes a triple-star system). However, the smaller blue star is mostly lost in the orange glare of its larger companion. While astronomers cannot physically see the individual stars, the light from those stars can tell astronomers a great deal about the binary system, including the speed at which the stars orbit one another and the ratio of the masses of the stars.

Like the Sun, the atmospheres of stars are composed predominantly of hydrogen and helium atoms, but they also contain an enormous array of other atoms, like carbon, oxygen, silicon, magnesium, and calcium. Each of those atoms has electrons which can be excited by absorbing photons emerging from deep within the star where temperatures are much higher. Depending on the composition and the temperature of the star's atmosphere, certain wavelengths of light will be absorbed by atoms. When astronomers study the number of photons of a certain energy and wavelength radiating from a star (referred to as the flux), they notice that the flux dips drastically at certain wavelengths. These are the spectral lines, caused by atoms in the star's atmosphere preferentially removing huge numbers of these photons.

For a star with a surface temperature of about 11,000 Kelvin, neutral helium atoms (He I) will be the primary absorbers of the star's photons, their electrons jumping to higher energy levels by absorbing some of the star's light. These atoms predominantly absorb photons with a wavelength of 4471 Å. Likewise, for a star with a surface temperature of about 4,500 Kelvin, ionized magnesium atoms (Mg II, which symbolizes a magnesium atom with one electron removed) will become excited by absorbing photons with a wavelength of 4481 Å.

Looking closely at stellar spectra, astronomers may notice two sets of superimposed spectral lines. The presence of both He I and Mg II is a strong sign that a single blob of light is, in fact, the combined light of two stars in a binary system: one with a surface temperature of 11,000 K and another with a temperature of 4,500 K. Therefore, even if the single individual stars cannot be seen, the tell-tale fingerprint of their individual spectra betrays the presence of two different stars. This is called a double-lined spectrum. These dominant spectral lines are constant for stars of 11,000 K and 4,500 K. The He I line in a stationary, unmoving star will always have a wavelength of 4471 Å. The Mg II line of an unmoving 4,500 K star will always be centered at 4481 Å. These unmoving wavelengths are called "rest wavelengths" and are symbolized by λ_0.

The motion of an object can distort the wavelength of absorbed or emitted photons, in much the same way as the motion of a car will distort the sound of its horn to a listener standing on the street. The sound of a car horn will be distorted by the car's approach (where the frequency or pitch of the horn will increase) or by the car's recession (where the frequency or pitch of the horn will decrease) for a listener. Likewise, if a light emitting-object is in motion relative to an observer – either approaching or receding – then the spectral lines will display a measurable shift, called either a blueshift when the star is approaching the Earth or a redshift when the star is receding. The shifting of a wavelength due to the relative motion between the object and the observer is called Doppler shift (see Unit 8.2).

Redshift refers to the movement of spectral lines toward longer wavelengths. The name redshift refers to the fact that red photons have the longest wavelength of the visible light photons. Redshifted velocities are designated as positive, since the star is adding distance between itself and the observer. The wavelength of the photon is elongated or stretched by the velocity of the star away from the observer. Blueshift refers to the movement of spectral lines toward shorter wavelengths. The wavelength of the photon is compressed by the motion of a star toward the observer. For a blueshifted light emitter, $\Delta\lambda$ and v_R will be negative, indicating that the star is subtracting distance between itself and the observer as it moves toward the Earth.

EQUATIONS AND CONSTANTS

Equation	Expression	Variables	
Doppler Shift	$\Delta\lambda = \lambda_{obs} - \lambda_0$	$\Delta\lambda$:	the Doppler shift
		λ_{obs}:	the wavelength of a shifted line
		λ_0:	the rest wavelength from an unmoving source
Radial Velocity	$v_R = \dfrac{\Delta\lambda \times c}{\lambda_0}$	v_R:	the radial velocity in km/ sec
		$\Delta\lambda$:	the Doppler shift
		c:	the speed of light in km/sec
		λ_0:	the rest wavelength from an unmoving source
Mass Ratio	$M = \dfrac{v_{r-max}}{v_{R-max}}$	M:	the mass ratio
		v_{r-max}:	the maximum velocity of the low mass binary star
		v_{R-max}:	the maximum velocity of the high mass binary star

ILLUSTRATIONS

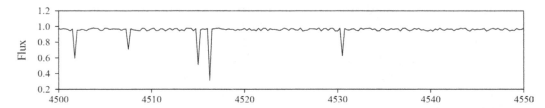

Figure 1. The spectrum of a star from 4500 Å to 4550 Å showing deep absorption lines from atoms in a single, hot star's atmosphere. The flux is the number of photons reaching Earth at a certain wavelength, with the spectral lines defined by sharp drops in the photon number. This is a single lined spectrum because all of the lines belong to atoms situated in a single star, with spectral lines at approximately 4502 Å, 4508 Å, 4515 Å, 4516 Å, and 4531 Å.

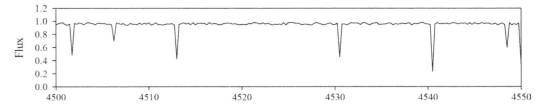

Figure 2. The spectra of a star from 4500 Å to 4550 Å showing deep absorption lines from atoms in a single, cool star's atmosphere. This is also a single lined spectrum.

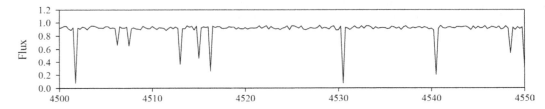

Figure 3. The spectrum of a star from 4500 Å to 4550 Å showing deep absorption lines from the combined light of both a cool and hot star's atmosphere. While the individual stars cannot be seen, their presence can be inferred by the double-lined nature of the spectrum.

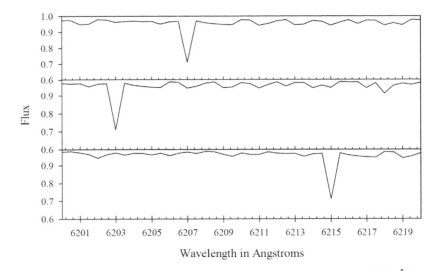

Figure 4. Three examples of absorption spectra showing an absorption line at rest at 6207 Å (top panel), blueshifted by -4Å to 6203 Å (middle panel) and redshifted by 8 Å to 6215 Å (bottom panel). These Doppler shifts of -4Å and +8Å correspond to radial velocities of -193 km/sec and +386 km/sec, respectively.

PROCEDURE

Real-world spectra from stars – mostly depending on their temperature – may have many thousands of spectral lines spread out over the entire visible spectrum. Datasheet #1 contains 21 evenly-spaced spectral images of a binary star system taken over the course of one complete orbit. The spectrum at each point has been magnified to displaying only the prominent He I and Mg II spectral lines indicative of a 11,000 K and a 4,500 K star, respectively. The spectral lines are caused by strong absorption of photons by electrons in magnesium and helium atoms, leading to a sharp drop in the flux of photons (number of photons reaching Earth) at those wavelengths. These spectral lines have well-measured rest wavelengths (λ_0) centered on 4471 Å and 4481 Å, respectively. The fluxes themselves – amounts of light at each wavelength – are inconsequential to the Doppler shift and the motion of the individual stars. The critical measurement is the wavelength of the spectral lines.

In Worksheet #1, for each of the given observations, carefully measure the wavelength of the He I and Mg II spectral lines. Motion of the stars toward or away from the Earth causes contractions or elongations of the absorbed photon wavelengths. Record the new, observed wavelengths (λ_{obs}) for each star. Calculate the Doppler shift ($\Delta\lambda$) for each phase and use the data to calculate the radial velocity of the stars at each point in orbit, making certain to use negative velocities for blue shifts, where appropriate.

In Worksheet #2, by using the radial velocities and phases, plot the changing motion of the stars over the course of one orbit. Plot the 11000 K star's velocity (the star displaying He I absorption) with a circle; plot the 4500 K star's velocity (the star displaying the Mg II absorption) with a square and connect the data points with a smooth, best-fit curve (do not simple connect the dots with straight line segments, as the motion of the stars in orbit is a smooth ellipse, not a sharp jump from position to position).

DATASHEET # 1

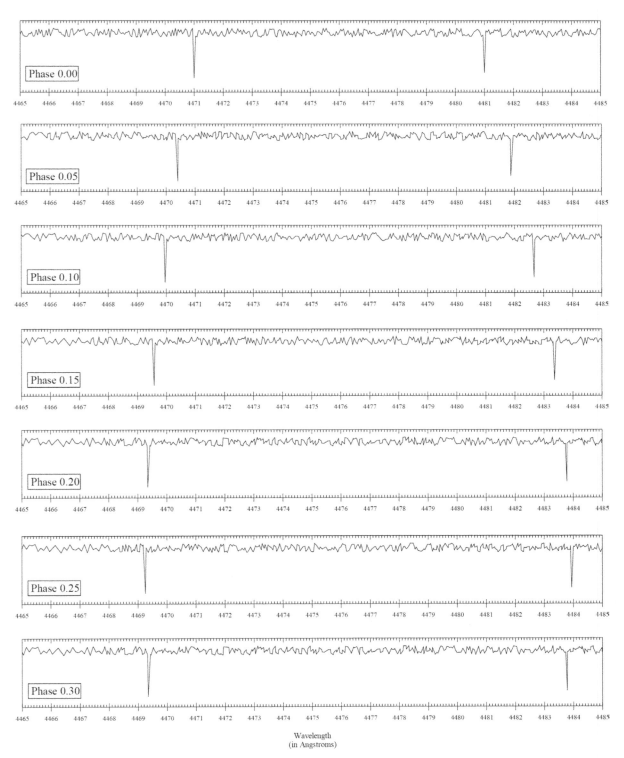

Wavelength
(in Angstroms)

Continue....

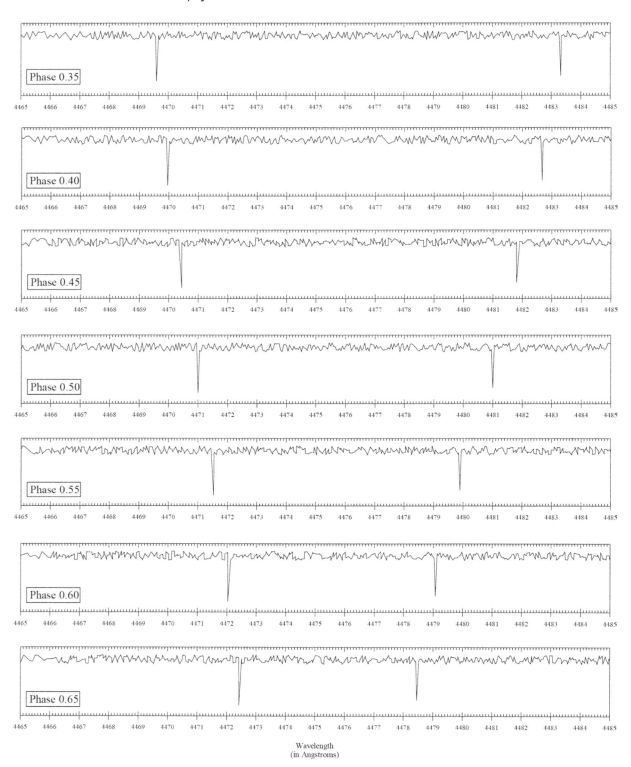

Wavelength
(in Angstroms)

Continue....

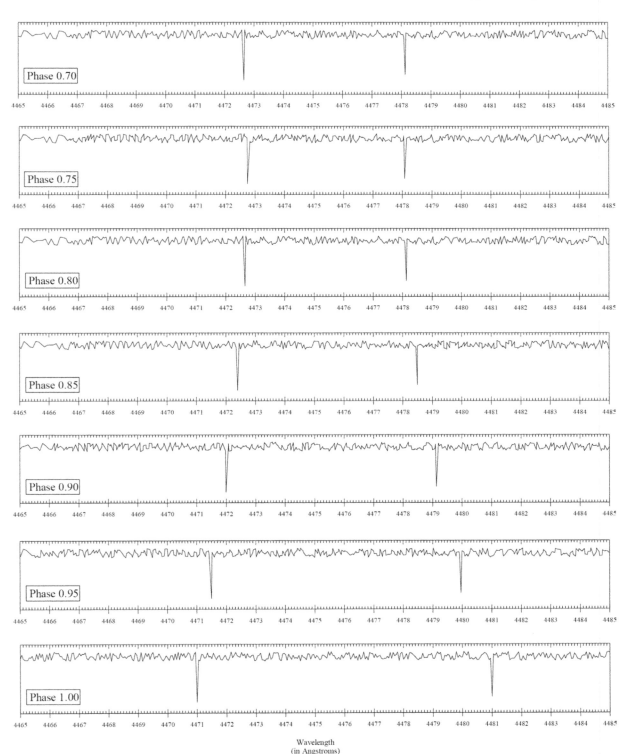

Wavelength
(in Angstroms)

Name _____ Id _____

Due Date _____ Lab Instructor _____ Section _____

Worksheet # 1

Phase	Measured He I Line (in Å)	Shift (Δλ) (in Å)	Radial Velocity (in km/sec)	Measured Mg II Line (in Å)	Shift (Δλ) (in Å)	Radial Velocity (in km/sec)
0.00						
0.05						
0.10						
0.15						
0.20						
0.25						
0.30						
0.35						
0.40						
0.45						
0.50						
0.55						
0.60						
0.65						
0.70						
0.75						
0.80						
0.85						
0.90						
0.95						
1.00						

1. What is the maximum radial velocity of the He I star?

2. What is the maximum radial velocity of the Mg II star?

3. What is the mass ratio of these stars?

4. Which star do you think is the more massive of the two? What led you to this conclusion?

Name _____ Id _____

Due Date _____ Lab Instructor _____ Section _____

Worksheet # 2

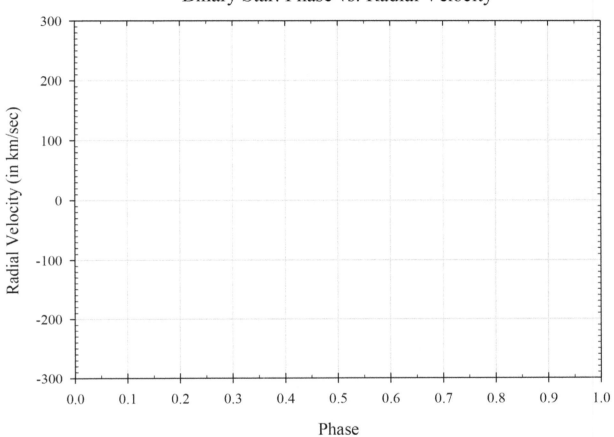

Binary Star: Phase vs. Radial Velocity

UNIT 8.5   ECLIPSING BINARIES

OBJECTIVE

To learn how astronomers can use the changing light output of a binary star system to measure the real properties of the stars, such as their masses and diameters. You will be able to read a light curve from a binary star system and pick out important points during the eclipse. This unit uses values of velocity and mass ratio from the spectroscopic binary studied in Unit 8.4.

INTRODUCTION

Half of the stars in the sky are binary stars, meaning that two (or more) stars formed from the same gas and dust cloud orbit one another in a repeating, periodic pattern. For some stars, the system is aligned in such a way that astronomers on Earth can see an eclipse: one star passing in front of another star. Much like a solar eclipse darkens the sky as the Moon covers the disk of the Sun, a stellar eclipse occurs when one star passes in front of another, blocking some or all of the other star's light as seen from the Earth.

The star Algol – translated as "head of the ogre" – is one of the most famous examples of eclipsing binary stars. This star is located in the constellation Perseus, located on the hip of the mythological hero. As the story goes, the hero Perseus killed and beheaded the monstrous Gorgon named Medusa, whose cursed stare could turn men to stone. Perseus carried the head with him to use as a weapon, as the enchanted gaze of the Gorgon still held its power. The star Algol reminded ancient sky watchers of Medusa's gaze, because it changed drastically in brightness every 3 days, dropping approximately 60% of its light output for several hours before returning to full brightness. For people who believed in the infinite, timeless, and unchanging nature of the heavens and stars, Algol presented an issue as it clearly changed. It was easier to believe that the star was the cursed eye of a mythological monster than accepting a challenge to the view of the nature of the Universe. Astronomers today realize that Algol is in fact an *eclipsing binary system:* two stars, almost perfectly aligned as seen from the Earth, which orbit one another and periodically block one another's light. One of the Algol stars is a cool, puffy, low-luminosity evolved star (called a "secondary" because of its lower luminosity) while the other star is a high-luminosity, high temperature main-sequence star (called the "primary" due to its high light output). Every three days, Algol will undergo two eclipses: the secondary blocks the light of the primary and then the primary blocks the light of the secondary. This cosmic carousel provides astronomers with a critical tool for measuring the characteristics of these two stars.

Eclipsing binary stars produce a light curve with tell-tale light changes that help astronomers to measure masses, periods, and sizes of stars. Primary first contact is the arrangement where – from the point of view of Earth – the stars are "touching" (see Figure 1). This is the last moment that their system is seen at its full brightness, with the light output of both stars unobstructed. Immediately after first contact, the eclipse begins. There is a quick and steady decline in light as the smaller secondary star begins to move in front of and block the light of the primary star. Second contact is the time when the secondary has reached a point where its entire disk is completely inside the disk of the primary (see Figure 2). This is the deepest point in the eclipse, since the maximum amount of light from the primary is being obscured by the secondary. Third contact is the beginning of the end of the eclipse. This is the last moment that the small star's disk appears completely surrounded by the primary star. At that moment, the eclipse breaks. Second-by-second the secondary slips out of alignment with the primary and – by fourth contact (see Figure 3) – the two stars are now unobscured again, returning the system to full brightness.

The orbit continues from this point, though we on Earth may not notice any changes in light output. The smaller star will whip around in its orbit, reaching a greatest elongation from the primary (the largest apparent distance between the two stars) before closing in on the larger star again. This time, however, with the smaller star on the far side of its orbit as seen from Earth, the secondary will slip completely behind the primary. Since the smaller, cooler star produces less light across its surface than the larger, hotter star, this secondary eclipse will be shallower than the primary eclipse. At secondary first contact, the small star and large star will appear side by side, as seen from Earth. At secondary second contact, the smaller star will pass entirely behind the larger star, with all of its light output blocked by the primary. The smaller star will remain hidden from sight until just after secondary third contact. At that moment, a sliver of the small star will peek out from behind that larger and begin adding its light to the system again. Finally, at secondary fourth contact, the small star will completely emerge and the system again will return to full brightness.

EQUATIONS AND CONSTANTS

Equation	Expression	Variables
Period-Velocity	$P = \dfrac{2\pi a}{v}$	P: period of the orbit, in sec a: the semi-major axis of the orbit, in km v: the velocity of the star, in km/sec
Total Mass	$(M + m) = \dfrac{a^3}{P^2}$	M: mass of the larger star, in units of $M_{\odot}$ m: mass of the smaller star, in units of $M_{\odot}$ P: the period of the orbit in years a: the semi-major axis (radius) of the orbit in AU
Stellar Radius (Primary Star)	$R = \dfrac{\pi a\left(T_4 - T_2\right)}{P}$	R: the radius of the large star, in km a: the semi-major axis of the orbit, in km T_2: the date of second contact T_4: the date of fourth contact P: the period of the orbit, in days
Stellar Radius (Secondary Star)	$r = \dfrac{\pi a\left(T_2 - T_1\right)}{P}$	r: the radius of the smaller star, in km a: the semi-major axis of the orbit, in km T_1: the date of first contact T_2: the date of second contact P: the period of the orbit, in days
Constants and Conversions		
1 day = 86,400 seconds		
1 year = 365.25 days		
1 AU = 1.496×10^8 km		

ILLUSTRATIONS

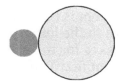

Figure 1. T_1 – First contact: This is the alignment of stars when the smaller star is just about to move in front of the larger star. The eclipse will begin immediately after this point as the smaller star moves across the disk of the larger star.

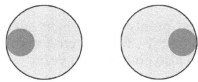

Figure 2. T_2 – Second contact: The smaller star is completely in front of the larger star. The eclipse reaches its deepest point as the small star blocks the light of the large, more luminous star. The small star will transit across the disk of the large star, maintaining the deep eclipse until T_3, or third contact, when the smaller star just aligns with the opposite limb of the larger star's disk. This is the last point of the full, deep eclipse.

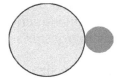

Figure 3. T_4 – Fourth contact: The eclipse ends as the star system returns to full brightness from the combined luminosity of both unobstructed stars. The smaller star will continue its orbit and reach a greatest western elongation (greatest separation to the right of the larger star) before heading back toward alignment. The secondary eclipse occurs as the smaller star is hidden behind that larger star. The secondary eclipse is not as deep as primary eclipse and lasts until the smaller star moves out of alignment and re-emerges from behind the larger star.

PROCEDURE

This lab is a follow up on the binary system covered in the previous unit featuring spectroscopic binaries. Using the Doppler shift, the orbital velocity of the secondary star was found to be 200 km/sec, with the orbital velocity of the primary falling at 120 km/sec. From those velocities, the mass ratio was determined as 1.667. These values were calculated from the Doppler shift of the stellar spectral lines.

Datasheet #1 shows the light output of a binary star system, composed of a large star with a surface temperature of 11,000 K (the primary star with the He I spectral absorption line) and a smaller star with a surface temperature of 4,500 K (the secondary star displaying Mg II spectral absorption). Out of eclipse (when the stars are not aligned), the stars have a combined apparent magnitude of approximately 5.4 (the jagged nature of the light output is due to the intrinsic flicker of the stars, as most stars have small temperature differences across their surfaces, as well as noise generated by the telescopic recording devices). When the stars align, as seen from the Earth, the brightness of the stellar system drops, as some of the light (or all of it, during a secondary eclipse) emerging from the surface of one star is blocked by the other star.

In a system like this one, where the two stars are perfectly aligned edge-on to the point of view of the Sun, astronomers on Earth will witness two periodic eclipses. Carefully mark off and measure the important points on the light curve: primary first contact (T_1), second contact (T_2), third contact (T$_3$), and fourth contact (T$_4$). Also, mark off the important points of the secondary eclipse: secondary first contact (t$_1$), second contact (t$_2$), third contact (t$_3$), and fourth contact (t$_4$).

Remembering that the period of a star system is the time it takes for the stars to travel from one configuration all through orbit and return to the initial alignment; also mark off the primary first contact of the next eclipse. The time it takes for the stars to orbit from T$_1$ to the following T$_1$ is the period, in days. Likewise, the time from first T$_2$ to the next T$_2$ passage is the same period. T$_3$ to T$_3$ would also yield the period. Any two identical points on the light curve would yield the period of the orbit, although some pairs of points are usually preferred for technical reasons. Clearly mark on your graph the two identical points you are using for the period.

Primary Eclipse: in the space provided on Worksheet #1, record your dates for T$_1$ through T$_4$.

Orbital Period: using your identical points on the light curve, measure the days which have passed between initial alignment (first T$_1$, first T$_2$, etc.) and second alignment (second T$_1$, T$_2$, etc.). Record that value in days and also convert it to years.

Orbital Velocities: these speeds are recalled from Unit 8.4 and were determined in km/sec. Convert these to km/day. Logically, given the number of seconds in one day, how far would these stars move in one day? Convert and record that value.

Semi-Major Axis: this is the radius of the orbit (in case of elliptical orbits: smallest value) and is dependent on two known parameters: the orbital period and the orbital speed. Rearranging the given Period-Velocity equation given in the Introduction, find the radius (a_1) of the large star's orbit and the radius (a_2) of the smaller secondary star's orbit.

Stellar Radius: knowing the times of eclipse, the size of the orbit, as well as the orbital period of the stars, determine the radius (in km) of each star. Make sure that your measurement units cancel out; this is done by using the same time-based units (an eclipse time in days and a period in days, for example).

DATASHEET # 1

Light Curve of Eclipsing Binaries

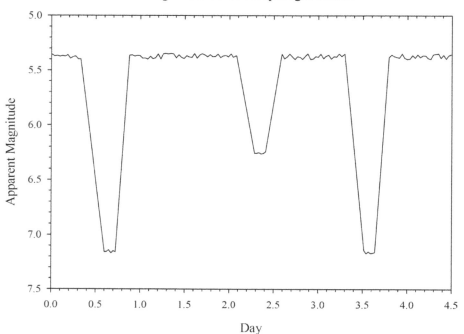

Name _____ Id _____

Due Date _____ Lab Instructor _____ Section _____

Worksheet # 1

Primary Eclipse Dates

First Contact T_1	Second Contact T_2	Third Contact T_3	Fourth Contact T_4

Orbital Period

Period (days)	Period (years)

Orbital Velocities

Primary Star (v_R) (in km/sec)	Primary Star (v_R) (in km/day)	Secondary Star (v_r) (in km/sec)	Secondary Star (v_r) (in km/day)
120		200	

Semi-Major Axis

Primary Star (a) (in km)	Primary Star (a) (in AU)	Secondary Star (a) (in km)	Secondary Star (a) (in AU)

Stellar Radius

Primary (R in km)	Secondary (r in km)

Total System Mass

$M + m$ (in $M_\odot$)

Stellar Masses

M (in $M_\odot$)	m (in $M_\odot$)

UNIT 8.6   MASSES OF EXOPLANET HOST STARS

OBJECTIVE

To learn how to compute stellar masses for planet-hosting stars based on the planetary semi-major axis of the planet's orbital period. Furthermore, learn to utilize the theoretical relationship between stellar masses and spectral types available for main-sequence stars (luminosity class V). Contemplate the differences between these two methods.

INTRODUCTION

Extrasolar planets or exoplanets are planets orbiting stars other than the Sun, implying that they are outside the Solar System. The existence of these objects helps us realize that there are many planets beyond Earth. At the present time, a large number of planets has been discovered via both ground-based facilities and space missions, notably the *Kepler* mission. As of May 8, 2016, over 2000 exoplanets have been found since 1988 (more specifically, 2125 planets in 1362 planetary systems, including 514 multiple planetary systems, have been confirmed).

Due to the vastness of this field of research, we are not in the position to convey a comprehensive summary and review on this topic. However, some important aspects stick out. For example, there are exoplanets that are much closer to their parent star than any planet in the Solar System is to the Sun. Mercury, the closest planet to the Sun, takes 88 days for an orbit, but the smallest known orbits of exoplanets have periods of only a few hours, i.e., Kepler-70b, as identified by S. Charpinet and team. Moreover, the Kepler-11 system, as identified by J. J. Lissauer and team, has five of its planets in smaller orbits than Mercury's. The High Accuracy Radial velocity Planet Searcher (HARPS) (since 2004) has discovered about a hundred exoplanets while the Kepler space telescope (since 2009) has already found more than a thousand. It has been found that there can be a significant number of planets per star. Additionally, a notable fraction of Sun-like stars is expected to have an "Earth-sized" planet in the habitable zone (see also Unit 6.6). Assuming 200 billion stars in the Milky Way, estimates indicate several billion potentially habitable Earth-sized planets in the Milky Way, rising to additional billions if planets orbiting the numerous red dwarfs (M-type stars) are included. Regarding G and K-type main-sequence stars, the most Earth-like planets (based on available information) include Kepler-452b (see J. M. Jenkins and team), and Kepler-442b (see G. Torres and team). Thus, exoplanets have been found for a large variety of stars, including main-sequence stars between spectral types A and M, as well as evolved stars.

There are also different types of exoplanets, though taking into account that the classification scheme is still under review. Examples include "Hot Jupiters" (i.e., gas giants close to their host stars that either formed extremely close-in or formed farther out and migrated inwardly), exo-Neptunes, and super-Earths. The latter objects are most likely rocky planets with masses greater than Earth, but no more than about 15 Earth-masses. In the Solar System all the planets but Mercury have near-circular orbits. However, many planets in extrasolar systems, especially massive planets with relatively long orbital periods, are often in highly eccentric orbits. But eccentric orbits are also found for some smaller planets; those cases are of interest regarding planet-planet interaction and planetary orbital instabilities, especially regarding densely-packed systems. Apparently, there may even by planets in retrograde orbits, as, e.g., ν Octantis, according to the work by D. J. Ramm and team. As time progresses, stay tuned for many more discoveries ahead.

Equations and Constants

Equation	Expression	Variables
Kepler's Third Law	$$M_\star = \frac{a^3}{P^2}$$	$M_\star$: the mass of the star, in units of $M_\odot$ a: the planet-star separation, in AU P: the orbital period of the planet, in years

Procedure

Part 1: Acquisition of Star-Planet System Data

Students are asked to compile data for the various star-planet systems, which are: planetary distances (semi-major axes), orbital periods, and stellar spectral types (see Worksheet #1, 2, and 3, as assigned). The planetary data and the stellar spectral types can be obtained from the website http://www.exoplanet.eu. In addition, students should also obtain the masses for the stars, referred to as M_* (from spectral type), based on the available information of Datasheet #1.

Part 2: Application of Kepler's Third Law

By focusing on the planet(s) of the selected star-planet systems, students are asked to record the planetary semi-major axes a and orbital periods P. This will allow computing the stellar masses based on Kepler's Third Law ("dynamic masses"). These masses should be recorded into the respective worksheet as (from Kepler's 3rd law).

Part 3: Comparison of Stellar Masses

Worksheet #4 is focused, among other tasks, on the comparison of the different types of stellar masses, and some statistical analysis.

Interpretation: After completion of Part 3, you will find that the stellar masses deduced by (1) employing the theoretical stellar mass—spectral type relationship and (2) applying Kepler's third law to the planetary orbits somewhat disagree. Does this violate the laws of physics? Of course not! Did you make any mistakes? Maybe, but let's assume not. So, how did this happen? In fact, there are two main reasons. Firstly, the theoretical stellar mass—spectral type relationship has been derived for main-sequence stars. However, even though all parent stars have previously been classified as "spectral-type V", in some cases, they have started to evolve away from the main-sequence, but their evolution has not sufficiently progressed to mandate a "spectral-type IV" classification. Secondly, even if the stars are still precisely spectral-type V, the theoretical stellar mass—spectral type relationship is only exactly valid for main-sequence "standard stars". In reality, however, stars are individuals, almost like people! They differ from the standard due to their distinct properties, such as: chemical composition, general fluctuations of surface temperatures, stellar rotation, magnetic fields, star spots (akin to sunspots), convective patterns, surface flows, etc. This also means that the "real" main-sequence has a certain width. In fact, there can be two main-sequence stars of identical spectral type and effective temperature, but slightly different luminosity!

DATASHEET # 1

Spectral Type	Temp (K)	Radius ($R_\odot$)	Mass ($M_\odot$)	Spectral Type	Temp (K)	Radius ($R_\odot$)	Mass ($M_\odot$)
F1 V	7042	1.541	1.56	G6 V	5603	0.937	0.888
F2 V	6909	1.480	1.52	G7 V	5546	0.923	0.863
F3 V	6780	1.453	1.48	G8 V	5486	0.910	0.838
F4 V	6653	1.427	1.44	G9 V	5388	0.870	0.814
F5 V	6528	1.400	1.40	K0 V	5282	0.830	0.790
F6 V	6403	1.333	1.33	K1 V	5169	0.790	0.766
F7 V	6280	1.267	1.26	K2 V	5055	0.750	0.742
F8 V	6160	1.200	1.19	K3 V	4973	0.730	0.718
F9 V	6047	1.155	1.12	K4 V	4730	0.685	0.694
G0 V	5943	1.120	1.05	K5 V	4487	0.640	0.670
G1 V	5872	1.100	1.02	K6 V	4294	0.601	0.643
G2 V	5800	1.000	1.00	K7 V	4133	0.565	0.614
G3 V	5754	0.990	0.967	K8 V	4006	0.533	0.582
G4 V	5708	0.980	0.940	K9 V	3911	0.505	0.547
G5 V	5657	0.950	0.914	M0 V	3850	0.480	0.510

Adapted from Astronomy and Astrophysics, Volume 526, id. A91, February 2011 by D.E. Fawzy and M. Cuntz.

Name _____ Id _____

Due Date _____ Lab Instructor _____ Section _____

Worksheet # 1

Stars With Planets

Star Name	Planet	Semi-major Planetary Axis a (AU)	Orbital Period P (years)	Spectral Type	$M_\star$ from spectral type ($M_\odot$)	$M_\star$ from Kepler's 3rd law ($M_\odot$)
51 Pegasi	b					
γ Cephei	b					
HD 179949	b					
τ Bootis	b					
HD 209458	b					
70 Virginis	b					
ε Eridani	b					
Gliese 876	b				0.334	
HD 8673	b					
HD 283750	b					
Kepler 22	b					
Kepler 186	f				0.48	
Kepler 438	b				0.54	
Kepler 442	b					
Kepler 452	b					

Name _____ Id _____

Due Date _____ Lab Instructor _____ Section _____

Worksheet # 2

Stars with Planetary Systems

Star Name	Planet	Semi-major Planetary Axis a (AU)	Orbital Period P (years)	Spectral Type	$M_\star$ from spectral type ($M_\odot$)	$M_\star$ from Kepler's 3rd law ($M_\odot$)
ups And A	b					
ups And A	c					
ups And A	d					
ups And A	e					
55 Cancri A	e					
55 Cancri A	b					
55 Cancri A	c					
55 Cancri A	f					
55 Cancri A	d					
HD 10180	b					
HD 10180	c					
HD 10180	i					
HD 10180	d					
HD 10180	e					
HD 10180	j					
HD 10180	f					
HD 10180	g					
HD 10180	h					

Name _____Id _____

Due Date _____ Lab Instructor _____ Section _____

Worksheet # 3

Stars – Planetary Systems of Choice

Star Name	Planet	Semi-major Planetary Axis a (AU)	Orbital Period P (years)	Spectral Type	$M_\star$ from spectral type $(M_\odot)$	$M_\star$ from Kepler's 3rd law $(M_\odot)$

Name _____ Id _____

Due Date _____ Lab Instructor _____ Section _____

Worksheet # 4

These post lab questions refer to Worksheet #1 and 2. Be prepared for additional questions based on Worksheet #3.

1. For the stars included in Worksheet #1, discuss the differences between the stellar masses based on the stellar spectral types and those obtained by Kepler's 3rd law. Which of those are considered more accurate? Why?

   (For stars with spectra types lower than M0V, Datasheet #1 is insufficient. Therefore, the results are already plugged in for your convenience.)

2. For the three planetary systems given in Worksheet #3, calculate the averages of masses obtained based on Kepler's 3rd law for each of the three systems. Which system shows the largest spread for these "dynamic" masses?

3. For the three planetary systems given in Worksheet #3, compare the averages of the "dynamic" masses with the respective values of masses obtained from the stellar spectral types. Discuss your results.

4. Repeat the previous the previous exercise for the star-planet system of HIP 105854; note that the star is of spectral type K2. Comment on the difference between the two methods of stellar mass determination. Also, check the luminosity class of the planet's host star.

UNIT 8.7   BLACK HOLES

OBJECTIVE

To learn about the nature and fundamentals of black holes, along with how astronomers measure their size. You will also be introduced to and use the concept of time dilation.

INTRODUCTION

Black holes are the natural evolutionary end to the life of a highly massive star. As the Universe's heaviest, brightest stars age and evolve into supergiants, their cores become incredibly hot and dense (reaching billions of degrees and densities hundreds of millions of times higher than naturally occurring materials, like gold or lead). As the cores become smaller and heavier, their own gravity becomes overwhelming. Stars use the energy of nuclear fusion to hold themselves up against their own immense mass and gravity, but eventually a star runs out of fuel and can no longer resist its own crushing internal weight. When the fuel is exhausted, the core buckles and collapses in on itself. Given the mass and density of such a core and the speed with which it collapses, there isn't any force in nature that can halt the collapse.

To the best of our understanding of current physics, without any force countering gravity, all of the mass of the entire core will collapse down to a microscopic, infinitely small, infinitely dense point in space. This is called a singularity, commonly known as a black hole: the mass of the collapsed core concentrated into zero volume (in a simplified understanding). Given that gravity is dependent on mass and the distance from that mass, the high mass and infinitely small size of the black hole means that its gravitational force becomes virtually infinite very close to the singularity. Black holes of this infinitely small size would not have a surface to speak of and would not be composed of ordinary, familiar matter.

An immediately obvious problem with the study of black holes is their gravitational pull. Matter and information are restricted to travel no faster than 300,000 km/sec: the speed of light. The escape velocity describes the speed needed to counter a massive body's downward gravitational pull and escape into space. The escape velocity depends on the mass of the central object and the distance from it. Very near the singularity, the escape velocity is near infinity. Further away from the singularity, the escape velocity may still be in the hundreds of millions or billions of kilometers per second, meaning that escape is impossible. The only hope of escape would be at a distance where the escape velocity is less than 300,000 km/sec. This distance is called the black hole's "event horizon."

The event horizon is not a tangible object or a part of the collapsed center black hole but instead is a region around the singularity where the escape velocity equals the speed of light. The mass of the singularity is the deciding factor for the event horizon distance. For a black hole with the mass of the Sun, the event horizon would be located approximately 3 kilometers from the singularity. For a black hole with a mass of 10 $M_\odot$, the event horizon would stretch 30 kilometers from the singularity. For a 1000 $M_\odot$ black hole, light could get no closer than 3000 kilometers before escape became impossible. Once matter, light, or information is closer to the 1000 $M_\odot$ black hole, it will be lost forever to the black hole. The distance from singularity to the event horizon is referred to as the Schwarzschild radius.

Near the black hole's event horizon, familiar aspects of reality become warped by the severe gravitational influence of the singularity. One of the most fascinating changes is called time dilation. While the passage of time may intuitively seem absolute and unchangeable, it is in fact highly influenced by gravity (if sufficiently strong), and two observers – one very near a strong gravity field and one very far from that gravity field – will measure different rates of time passage. This is not due to mechanical issues with the clocks but with the actual bending of space itself and a change in the way time passes in that space. Closer to a strong gravitational influence, the passage of time appears to pass more slowly to an outside observer. For example, extremely distant satellites – far from the gravity well of the Earth – have clocks that run 45 microseconds slower than Earth-based clocks per day. While a small effect, it is measureable and present.

This time dilation effect can become incredibly prevalent near an extreme gravity field; namely, a black hole's event horizon. The closer a clock is placed to an event horizon, the slower its time will pass to an outside observer. For example, if the Sun was replaced with a 5 million solar mass, 3 million kilometer wide black hole (about the same size as the black hole located in the heart of the Milky Way galaxy), time on a Black-Hole orbiting Earth would tick by 0.5% slower than on a Sun-orbiting Earth. The passage of 24 hours on a Sun-orbiting Earth would register as only 23.88 hours (23 hours 53 seconds) on the Black-Hole orbiting Earth. Closer and closer to the event horizon, time would pass slower and slower. 14 million kilometers from the event horizon, 24 standard hours would only register as 22 hours and 46 minutes to someone orbiting the black hole. 140,000 kilometers from the event horizon and 24 hours to the outside world would only register as 7 hours on the black hole's clock. For an observer who was just 140 kilometers from the event horizon, 24 Sun-orbiting Earth hours would pass in only 14 minutes. Spending 30 days orbiting close to the event horizon would see the outside Universe age a decade.

Inside the event horizon, all matter and energy is funneled down into the singularity with astonishing speed and force. Even moving away from the singularity is impossible, since backing up even one nanometer (or less) would require a speed greater than the speed of light which (based on our understanding) is not physically possible. Therefore, anything crossing the event horizon is fated to be driven into the singularity and compressed down into the black hole's mass.

The actual make-up of the singularity and the matter that forms it may always remain a mystery to scientists, as information cannot escape from the black hole. The more matter piled into the black hole, the more massive it becomes, with the event horizon swelling outward to encompass more and more volume of space. Every one solar mass added to the black hole swells its event horizon by three kilometers. Eventually – if a black hole consumes enough material and integrates the captured mass into the singularity – the mass and event horizon will bulge to giant proportions. These are known as supermassive black holes, some one of the largest of which – residing at the center of the largest galaxy of the extremely distant, densely packed Pegasus galaxy cluster – tips the scales at 20 billion times the mass of the Sun with an event horizon over 100 billion kilometers across. While these may sound like outrageously exotic, rare objects, almost every spiral galaxy observed so far is host to a supermassive black hole in its core.

Finally, let us consider some aspects about the types of black holes as observed and relevant to astronomy. Originally, stellar black holes have been proposed as a consequence of the collapse of supermassive stars. The upper mass limit of stellar black holes is about 25 to 30 $M_\odot$. It is set by the maximal initial stellar masses on the main-sequence, approximately given as 120 to 150 $M_\odot$, and the property that these stars will lose 80 to 90% of their initial mass during stellar evolution. Supermassive black holes, discovered to reside at the center of spiral galaxies have masses between 10^6 and 10^{10} $M_\odot$; this type of black holes is formed in the early stages of the Universe, and are considered relevant for the formation and dynamics of galaxies. Another type of black holes is called intermediate-mass black holes with masses between 100 and 10^6 $M_\odot$. However, the origin of this type of black holes is not yet understood by scientists.

EQUATIONS AND CONSTANTS

Equation	Expression	Variables
Schwarzschild Radius	$R_S = \dfrac{2GM}{c^2}$	G: the gravitational constant M: black hole's mass in kg c: speed of light in m/s
Simplified Schwarzschild Radius	$R_S = 2.949 \times M_{BH}$	R_S: the radius of the event horizon in km M_{BH}: black hole's mass in solar masses
Time Dilation	$t_{BH} = t_{OBH}\sqrt{1 - \dfrac{R_S}{R}}$	t_{BH}: the time according to an observer near a black hole t_{OBH}: the time according to an observer very far away from a black hole R_S: the Schwarzschild radius of the black hole R: the distance from the center of the black hole (in the same length unit at R_s)

Constants and Conversions

$$1\ M_\odot = 1.98895 \times 10^{30}\ \text{kg}$$

$$1\ \text{AU} = 1.496 \times 10^8\ \text{km}$$

$$c = 299{,}792.458\ \frac{\text{km}}{\text{s}} \approx 300{,}000\ \frac{\text{km}}{\text{s}}$$

$$G = 6.67408 \times 10^{-11}\ \frac{\text{m}^3}{\text{kg}\ \text{s}^2}$$

PROCEDURE

Worksheet #1 is composed of two tables with data on some well-studied and important black holes. Table 1 contains the calculated masses of 5 stellar black holes; those black holes formed from the collapse of the cores of massive stars. The list includes the first black hole discovered (Cygnus X-1) along with one of the heaviest stellar black holes (M33 X7), the smallest black hole currently known (XTE J1650-500) as well as two very active, matter-accreting black holes. Using the simplified Schwarzschild equation and the given masses, find the radius of the event horizon (in km). Table 2 contains data on four supermassive black holes, including the heaviest known black hole, the Milky Way's black hole, and the quasar-producing black hole 3C 327. Find their event horizon radii. The Schwarzschild radius equation will produce a radius in kilometers; convert that radius in AU and record that resulting supermassive black hole size in the table. Notice that since these black holes are extremely large; their radii are best expressed in AU, the measurement usually reserved for the size of planetary orbits.

Worksheet #2 has a schematic outline of the New York metropolitan region, centered on Manhattan with Long Island to the east (right) and New Jersey on the west (left). On this map, sketch the largest and smallest stellar mass black hole considered (M33 X7 and XTE J1650, respectively). The line labeled "10 km" on the map represents a distance of 10 km in the real world. Despite having masses many millions of times that of planet Earth, even the largest stellar black hole could easily fit into a city-sized area!

Pick a spot on the map near the center and mark it with a small **x** to represent the center of the black hole (the singularity). Using the scale given on the map and the Schwarzschild radii calculated in Table 1, mark the distances to the event horizon. With a compass, draw a full circle around the singularity to represent the full diameter of the event horizon. Repeat this process for the second black hole. Most of the known stellar black holes fall between these two sizes.

Finally, Worksheet #3 deals with one of the more unique and fascinating properties of black holes: time dilation near the event horizon. Using the time dilation equation, determine how the passage of time is warped as a consequence of the proximity to a black hole. Remember that the variable R_s is the distance from singularity to event horizon, with R the distance between the singularity and an observer orbiting near the black hole. The closer R and R_s get to one another, the slower time appears to pass to the outside Universe.

Name _____ Id _____

Due Date _____ Lab Instructor _____ Section _____

Worksheet # 1

Table 1

Black Hole Name	Black Hole Mass $(M_\odot)$	Schwarzschild Radius (km)
Cygnus X-1	15.0	
M33 X-7	15.7	
XTE J1650-500	3.8	
V404 Cygni	12.5	
V1487 Aquilae	10.0	

Table 2

Black Hole Name	Host Galaxy	Black Hole Mass $(M_\odot)$	Schwarzschild Radius (AU)
Sagittarius A*	Milky Way	4.2×10^6	
P3	Andromeda	1.4×10^8	
3C 273	Quasar 3C 273	9.6×10^8	
Phoenix SMBH	Phoenix Cluster	2.0×10^{10}	

Data adopted from Wikipedia

Name _____ Id _____

Due Date _____ Lab Instructor _____ Section _____

Worksheet # 2

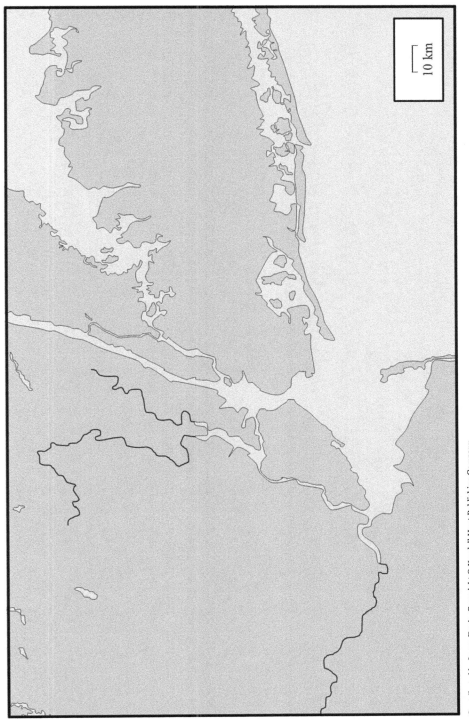

Name _____ Id _____

Due Date _____ Lab Instructor _____ Section _____

Worksheet # 3

For the following questions, consider the black hole in question to be Centaurus A, an active black hole at the heart of galaxy NGC 5128 with a diameter similar to that of the Earth's orbit around the Sun.

1. If assumed as perfectly circular, the Earth's orbit has a radius of 1.495978707×10^8 km (or exactly 1 AU). This is on par with the size of the Schwarzschild radius of some supermassive black holes. What would be the mass (in $M_\odot$) of a black hole with a Schwarzschild radius of 1 AU?

2. An accurate clock far, far from any black hole keeps track of the passage of time. Most of the Universe agrees that after 1 year, 365.25 Earth-days have passed. An identical clock sits on the surface of a planet orbiting around a supermassive black hole with a distance from the singularity of 2.0 AU. How much time has passed according to the black hole orbiting clock?

3. The International Space Station orbits 400 kilometers above the surface of the Earth. If a space station was placed in orbit 400 kilometers above the black hole's event horizon, how much time would pass on its clock while 365.25 days pass on Earth?

4. Imagine you fly toward this black hole at very high speed, establishing an orbit of 100 km from the event horizon. You orbit there for 7 days before leaving and heading back to Earth. How much time passed in the Universe during your 7 days in orbit?

TABLES

Table 1: Astronomical and Physical Constants

1. Astronomical Constants	
Astronomical Unit	$= 1.496 \times 10^{11}$ m
Parsec	$= 3.086 \times 10^{16}$ m
	$= 3.262$ light-years
Light-year	$= 9.461 \times 10^{15}$ m
	$= 63{,}240$ AU
Mass of Sun	$= 1.989 \times 10^{30}$ kg
Mass of Earth	$= 5.974 \times 10^{24}$ kg
Mass of Moon	$= 7.348 \times 10^{22}$ kg
Radius of Sun	$= 6.960 \times 10^{8}$ m
Radius of Earth (Equator)	$= 6.378 \times 10^{6}$ m
Radius of Moon (Equator)	$= 1.738 \times 10^{6}$ m
2. Physical Constants	
Gravitational Constant	$= 6.674 \times 10^{-11}$ N m^2 kg^{-2}
Speed of Light	$= 2.998 \times 10^{8}$ m s^{-1}
Boltzmann's Constant	$= 1.381 \times 10^{-23}$ J K^{-1}
Planck's Constant	$= 6.626 \times 10^{-34}$ J s
Stefan-Boltzmann Constant	$= 5.670 \times 10^{-8}$ W m^{-2} K^{-4}
Wien's Constant	$= 2.898 \times 10^{6}$ nm K

Table 2: The Eight Planets Planetary Orbital Parameters

Planetary Orbital Characteristics

Planet	Semi-major axis (AU)	Sidereal period (years)	Synodic period (days)	Orbital eccentricity	Inclination of orbit to ecliptic (°)
Mercury	0.3871	0.2408	115.88	0.206	7.00
Venus	0.7233	0.6152	583.92	0.007	3.39
Earth	1.0000	1.0000	…	0.017	0.00
Mars	1.5237	1.8809	779.94	0.093	1.85
Jupiter	5.2034	11.862	398.9	0.048	1.31
Saturn	9.5371	29.458	378.1	0.054	2.49
Uranus	19.1913	84.01	369.7	0.047	0.77
Neptune	30.0690	164.79	367.5	0.009	1.77

Planetary Physical Characteristics

Planet	Diameter at equator (km)	Mass (kg)	Rotation period (days)	Inclination of equator to orbit (°)	Albedo	Number of moons*
Mercury	4,879	3.302×10^{23}	58.646	0.0	0.106	0
Venus	12,104	4.869×10^{24}	−243.01	177.3	0.65	0
Earth	12,756	5.974×10^{24}	0.997	23.45	0.37	1
Mars	6,794	6.419×10^{23}	1.026	25.19	0.15	2
Jupiter	142,984	1.899×10^{27}	0.414	3.12	0.52	67
Saturn	120,536	5.685×10^{26}	0.444	26.73	0.47	62
Uranus	51,118	8.685×10^{25}	−0.720	97.86	0.50	27
Neptune	49,528	1.024×10^{26}	0.671	29.56	0.50	13

− indicated that the rotation is retrograde (opposite to the planet's orbital motion)
* as of Dec 1, 2015

Table 3: Kuiper Belt Objects

Name	Original Name	a (AU)	e	Incl. (°)	Diameter[*] (km)	Discovery	Year
Pluto[**]	...	39.541	0.249	17.1	2375	C. Tombaugh	1930
Eris[**]	2003 UB_{313}	67.781	0.440	44.0	2326	M. Brown, C. Trujillo, D. Rabinowitz	2005
Haumea[†]	2003 EL_{61}	43.218	0.191	28.2	1900 × 990	F. Aceituno, P. Santos-Sanz, J. Ortiz	2005
Makemake	2005 FY_9	45.660	0.156	29.0	1440	M. Brown, C. Trujillo, D. Rabinowitz	2005
...	2007 OR_{10}	66.850	0.506	30.9	1300	M. Brown, M. Schwamb, D. Rabinowitz	2007
Charon[+]	...	...	...	...	1210	J. Christy	1978
Ixion	2001 KX_{76}	39.680	0.242	19.6	1200	Deep Elliptic Survey	2001
Quaoar	2002 LM_{60}	43.405	0.039	8.0	1110	C. Trujillo, M. Brown	2002
Orcus	2004 DW	39.343	0.227	20.6	1100	M. Brown, C. Trujillo, D. Rabinowitz	2004
Sedna	2003 VB_{12}	524	0.855	11.9	1000	M. Brown, C. Trujillo	2003
Varuna	2000 WR_{106}	42.904	0.056	17.2	900	Spacewatch	2000

[*] Except for Pluto and Charon, the diameters as listed are somewhat uncertain

[**] Between 1930 and 2006, Pluto had the status of a planet. When Eris was discovered, it was thought to be larger in size than Pluto, a finding that was revised when more accurate measurements of both Pluto and Eris became available. However, Eris is found to be more massive than Pluto due to its higher density.

[+] Moon of Pluto

[†] Haumea is highly elongated, with an equatorial diameter (1900 km) far larger than its pole-to-pole diameter (900 km).

Table 4: The Nearest Stars

Name	Spectral Class	M	m	Distance (pc)	Luminosity Class	Color (B-V)
Sun	G2	4.83	−26.7	4.85×10^{-6}	V	0.65
Proxima Centauri	M5	15.6	11.1	1.30	V	1.82
Alpha Centauri A	G2	4.38	0.01	1.34	V	0.71
Alpha Centauri B	K1	5.70	1.33	1.34	V	0.88
Barnard's Star	M4	13.2	9.51	1.83	V	1.71
Wolf 359	M6.5	16.6	13.54	2.41	V	2.03
Lalande 21185	M2	10.5	7.52	2.55	V	1.44
Luyten 726-8 A	M6	15.6	12.7	2.68	V	1.87
Luyten 726-8 B	M6	16.1	13.2	2.68	V	1.98
Sirius A	A1	1.42	−1.46	2.64	V	0.00
Sirius B	wd* (B1)	11.2	8.44	2.64	wd	−0.20
Ross 154	M3.5	13.1	10.44	2.94	V	1.76
Ross 248	M5.5	14.8	12.3	3.16	V	1.92
Epsilon Eridani	K2	6.19	3.74	3.21	V	0.89
Ross 128	M4	13.5	11.13	3.35	V	1.59
61 Cygni A	K5	7.51	5.20	3.50	V	1.14
61 Cygni B	K7	8.22	6.05	3.50	V	1.32
EZ Aquarii	M5	15.6	13.3	3.40	V	1.96
Procyon A	F5	2.66	0.34	3.51	V/IV	0.40
Procyon B	wd* (A1)	13.0	10.7	3.51	wd	0.20
Epsilon Indi	K5	6.9	4.83	3.62	V	1.06
DX Cancri	M6.5	17.0	14.81	3.63	V	2.08
Groombridge 34 A	M1.5	10.3	8.09	3.59	V	1.56
Groombridge 34 B	M3.5	13.3	11.06	3.59	V	1.80
Tau Ceti	G8.5	5.69	3.50	3.65	V	0.72
Lacaille 9352	M0.5	9.8	7.34	3.28	V	1.50
Luyten's Star	M4	11.9	9.87	3.74	V	1.57
Lacaille 8760	M0	8.69	6.67	3.95	V	1.40
Kruger 60 A	M3	11.8	9.59	4.04	V	1.65
Kruger 60 B	M4	13.5	11.40	4.04	V	1.80

Continue....

Kapteyn's Star	M1	10.9	8.85	3.91	V	1.57
Ross 614 A	M4.5	13.3	11.1	4.10	V	1.72
Ross 614 B	M8	16.8	14.2	4.10	V	...
M = absolute magnitude, m = apparent magnitude						

* This object is classified as a white dwarf. The spectral class is given for tutorial reasons only; it is an equivalent based on its surface temperature.
Note: All listed stars are main-sequence stars with the exception of Sirius B, Procyon A and B. Furthermore this table does not consider brown dwarfs, which are not stars.

Table 5: The Brightest Stars

Star	Name	Spectral Class	M	m	Distance (pc)	Luminosity Class	Color (B-V)
Sun		G2	4.83	−26.7	4.85×10^{-6}	V	0.65
α CMa	Sirius	A1	1.42	−1.46	2.64	V	0.00
α Car	Canopus	F0	−5.65	−0.74	96	I/II	0.15
α Cen	Rigil Kent	G2	4.38	0.01	1.34	V	0.71
α Boo	Arcturus	K0	−0.30	−0.05	11.26	III	1.23
α Lyr	Vega	A0	0.58	0.03	7.68	V	0.00
α Aur	Capella Aa	K0	0.35	0.08	13.12	III	0.80
α Aur	Capella Ab	G1	0.20	0.16	13.12	III	0.65
β Ori	Rigel	B8	−7.92	0.13	260	I	−0.03
α CMi	Procyon	F5	2.66	0.34	3.51	V/IV	0.40
α Ori	Betelgeuse	M2	−5.85	0.42	197	I	1.85
α Eri	Achernar	B6	−2.77	0.46	43	V	−0.16
β Cen	Hadar	B1	−4.53	0.61	120	III	−0.23
α Aql	Altair	A7	2.22	0.76	5.13	V	0.22
α Cru		B1	−4.14	0.77	99	IV	−0.23
α Tau	Aldebaran	K5	−0.64	0.86	20	III	1.54
α Sco	Antares	M1	−5.28	0.96	170	I	1.83
α Vir	Spica	B1	−3.5	0.97	77	V	−0.23
β Gem	Pollux	K0	1.08	1.14	10.4	III	1.00
α PsA	Fomalhaut	A3	1.72	1.16	7.61	V	0.09
α Cyg	Deneb	A2	−8.38	1.25	802	I	0.09
β Cru		B0.5	−4.6	1.25	280	III	−0.23
α Leo	Regulus	B8	−0.52	1.40	24.3	IV	−0.11
α CMa	Adhara	B2	−4.8	1.50	430	II	−0.21
α Gem	Castor	A1	0.99	1.93	51	V	0.03
λ Sco	Shaula	B2	−3.70	1.62	180	IV	−0.24
γ Ori	Bellatrix	B2	−2.79	1.64	77	III	−0.21
β Tau	Elnath	B7	−1.34	1.65	40	III	−0.13

Continue....

Table 6: Variable Stars

Type	Spectral Class	Period (days)	Average Absolute Magnitude (M)	Change in Magnitude (ΔM)	Example
Cepheids (Type I)	F and G	3 to 50	−1.5 to −5	0.1 to 2	Delta Cep
Cepheids (Type II)	F and G	5 to 30	0 to −3.5	0.1 to 2	W Vir
RR Lyrae	A to F	hours	0 to −1	1 to 2	RR Lyr
Beta Cephei	B	0.1 to 0.3	−2 to −4	0.1	Beta Cep
Delta Scuti	F	hours	0 to −2	0.25	Delta Sct
Mira Type	M	80 to 600	−2 to −2	2.5	Omicron Cet (Mira)
Semiregular Variable	M	30 to 2000	0 to −3	1 to 2	Alpha Ori (Betelgeuse)
Irregular	All	Irregular	near 0	0 to 4	Pi Gru
RV Tauri	G to K	30 to 150	−2 to −3	0 to 3	RV Tau
Spectrum-Variable	A	1 to 25	0 to −1	0.1	

Table 7: The Messier Catalogue

Between 1771 and 1786 the comet hunter Charles Messier and his colleague Pierre Mechain published a list of 109 objects that could be mistaken for comets. From this M 40, M 73, M 91 and M 102 have been omitted. M 40 is a double star, M 73 has only 4 stars, M 91 has not been observed by others, and M 102 is probably a repetition of M 101.

Many of the other 105 objects may be observed with binoculars, most can be seen with a 2½ to 3 inch telescope, and some amateurs have reported finding every object in the catalogue with a 6-inch telescope.

M	NGC	RA	Dec	mag	Constellation	Description	Date Observed
1	1952	05h31m	+22°	9	Taurus	Nebula	
2	7089	21h31m	−1°	7	Aquarius	Globular Cluster	
3	5272	13h40m	+28°	6	Canes Venatici	Globular Cluster	
4	6121	16h22m	−26°	6	Scorpius	Globular Cluster	
5	5904	15h17m	+2°	6	Serpens	Globular Cluster	
6	6405	17h39m	−32°	6	Scorpius	Open Cluster	
7	6475	17h53m	−35°	5	Scorpius	Open Cluster	
8	6523	18h2m	−24°		Sagittarius	Nebula	
9	6333	17h18m	−19°	8	Ophiuchus	Globular Cluster	
10	6254	16h56m	−4°	6	Ophiuchus	Globular Cluster	
11	6705	18h50m	−6°	7	Scutum	Open Cluster	
12	6218	16h46m	−2°	7	Ophiuchus	Globular Cluster	
13	6205	16h41m	+37°	6	Hercules	Globular Cluster	
14	6402	17h37m	−3°	8	Ophiuchus	Globular Cluster	
15	7078	21h29m	+12°	6	Pegasus	Globular Cluster	
16	6611	18h18m	−14°	7	Serpens	Open Cluster	
17	6618	18h20m	−16°	7	Sagittarius	Nebula	
18	6613	18h19m	−17°	7	Sagittarius	Open Cluster	
19	6273	17h01m	−26°	7	Ophiuchus	Globular Cluster	
20	6514	18h01m	−23°		Sagittarius	Nebula	
21	6531	18h03m	−23°	7	Sagittarius	Open Cluster	
22	6656	18h35m	−24°	5	Sagittarius	Globular Cluster	
23	6494	17h56m	−19°	6	Sagittarius	Open Cluster	
24	6603	18h17m	−18°	6	Sagittarius	Open Cluster	

Continue....

25 IC	4725	18h31m	−19°	6	Sagittarius	Open Cluster	
26	6694	18h44m	−9°	9	Scutum	Open Cluster	
27	6853	19h59m	+23°	8	Vulpecula	Nebula	
28	6626	18h23m	−25°	7	Sagittarius	Globular Cluster	
29	6913	20h23m	+38°	8	Cygnus	Open Cluster	
30	7099	21h39m	−23°	8	Capricornus	Globular Cluster	
31	224	00h42m	+41°	4	Andromeda	Galaxy	
32	221	00h42m	+41°	9	Andromeda	Galaxy	
33	598	01h33m	+31°	6	Triangulum	Galaxy	
34	1039	02h41m	+43°	6	Perseus	Open Cluster	
35	2168	06h08m	+24°	6	Gemini	Open Cluster	
36	1960	05h35m	+34°	6	Auriga	Open Cluster	
37	2099	05h55m	+33°	6	Auriga	Open Cluster	
38	1912	05h27m	+36°	6	Auriga	Open Cluster	
39	7092	21h32m	+48°	6	Cygnus	Open Cluster	
41	2287	06h46m	−21°	6	Canis Major	Open Cluster	
42	1976	05h34m	−6°		Orion	Nebula	
43	1982	05h35m	−5°		Orion	Nebula	
44	2632	08h40m	+20°	4	Cancer	Open Cluster	
45	—	03h46m	+24°	2	Taurus	Open Cluster	
46	2437	07h42m	−15°	6	Puppis	Open Cluster	
47	2422	07h36m	−14°	5	Puppis	Open Cluster	
48	2548	08h13m	−6°	6	Hydra	Open Cluster	
49	4472	12h29m	+8°	9	Virgo	Galaxy	
50	2323	07h02m	−8°	7	Monoceros	Open Cluster	
51	5194	13h29m	+47°	8	Canes Venatici	Galaxy	
52	7654	23h23m	+61°	7	Cassiopeia	Open Cluster	
53	5024	13h12m	+18°	8	Coma Berenices	Globular Cluster	
54	6715	18h54m	−31°	8	Sagittarius	Globular Cluster	
55	6809	19h39m	−31°	6	Sagittarius	Globular Cluster	
56	6779	19h16m	+30°	8	Lyra	Globular Cluster	
57	6720	18h53m	+33°	9	Lyra	Nebula	
58	4579	12h37m	+12°	10	Virgo	Galaxy	

Continue....

59	4621	12h41m	+12°	10	Virgo	Galaxy	
60	4649	12h43m	+12°	9	Virgo	Galaxy	
61	4303	12h21m	+5°	10	Virgo	Galaxy	
62	6266	17h00m	−30°	7	Scorpius	Globular Cluster	
63	5055	13h15m	+42°	9	Canes Venantici	Galaxy	
64	4826	12h56m	+22°	9	Coma Berenices	Galaxy	
65	3623	11h18m	+13°	10	Leo	Galaxy	
66	3627	11h19m	+13°	9	Leo	Galaxy	
67	2682	08h50m	+12°	7	Cancer	Open Cluster	
68	4590	12h38m	−27°	8	Hydra	Globular Cluster	
69	6637	18h30m	−32°	8	Sagittarius	Globular Cluster	
70	6681	18h42m	−32°	8	Sagittarius	Globular Cluster	
71	6838	19h53m	+19°	7	Sagittarius	Globular Cluster	
72	6981	20h52m	−13°	9	Aquarius	Globular Cluster	
74	628	01h36m	+16°	10	Pisces	Galaxy	
75	6864	20h05m	−22°	8	Sagittarius	Globular Cluster	
76	650	01h41m	+51°	11	Perseus	Nebula	
77	1068	02h42m	0°	9	Cetus	Galaxy	
78	2068	05h46m	0°		Orion	Nebula	
79	1904	05h23m	−25°	7	Lepus	Globular Cluster	
80	6093	16h16m	−23°	7	Scorpius	Globular Cluster	
81	3031	09h54m	+69°	7	Ursa Major	Galaxy	
82	3034	09h54m	+70°	9	Ursa Major	Galaxy	
83	5236	13h36m	−30°	8	Hydra	Galaxy	
84	4374	12h24m	+13°	10	Virgo	Galaxy	
85	4382	12h24m	+18°	10	Coma Berenices	Galaxy	
86	4406	12h25m	+13°	10	Virgo	Galaxy	
87	4486	12h30m	+13°	9	Virgo	Galaxy	
88	4501	12h31m	+15°	10	Coma Berenices	Galaxy	
89	4552	12h35m	+13°	10	Virgo	Galaxy	
90	4569	12h36m	+13°	10	Virgo	Galaxy	
91	4548	12h35m	+14°	11	Coma Berenices	Galaxy	

Continue....

92	6341	17h17m	+43°	6	Hercules	Globular Cluster	
93	2447	07h44m	−24°	6	Puppis	Open Cluster	
94	4736	12h50m	+41°	8	Canes Venatici	Galaxy	
95	3351	10h43m	+12°	10	Leo	Galaxy	
96	3368	10h46m	+12°	10	Leo	Galaxy	
97	3587	11h14m	+55°	11	Ursa Major	Nebula	
98	4192	12h13m	+15°	10	Coma Berenices	Galaxy	
99	4254	12h18m	+15°	10	Coma Berenices	Galaxy	
100	4321	12h22m	+16°	10	Coma Berenices	Galaxy	
101	5457	14h03m	+54°	8	Ursa Major	Galaxy	
103	581	01h32m	+61°	7	Cassiopeia	Open Cluster	
104	4594	12h39m	−12°	8	Virgo	Nebula	
105	3379	10h47m	+13°	10	Leo	Galaxy	
106	4258	12h18m	+47°	9	Canes Venatici	Galaxy	
107	6171	16h32m	−13°	9	Ophiuchus	Globular Cluster	
108	3556	11h11m	+56°	11	Ursa Major	Galaxy	
109	3992	11h57m	+54°	11	Ursa Major	Galaxy	

All coordinates are approximately the values for 2015, all magnitudes are apparent, and are rounded to the nearest whole number.

NAMED MESSIER OBJECTS

M1   Crab Nebula
M8   Lagoon Nebula
M13   Hercules Cluster
M17   Omega Nebula
M20   Trifid Nebula
M27   Dumbbell Nebula
M31   Andromeda Galaxy
M33   Triangulum Galaxy
M42   Orion Nebula
M44   Beehive Cluster
M45   Pleiades
M51   Whirlpool Galaxy
M57   Ring Nebula
M63   Sunflower Galaxy
M97   Owl Nebula
M101   Pinwheel Galaxy
M104   Sombrero Nebula

Table 8: Coordinates of the Ecliptic

RA (hr)	RA (°)	Dec (°)
0	360	0
23	345	−6
22	330	−12
21	315	−17
20	300	−20
19	285	−23
18	270	−23.5
17	255	−23
16	240	−20
15	225	−17
14	210	−12
13	195	−6
12	180	0
11	165	6
10	150	12
9	135	17
8	120	21
7	105	22
6	90	23.5
5	75	22
4	60	21
3	45	17
2	30	12
1	15	6

(The declination has been rounded to the nearest degree)

Source Information and Suggested Readings

Text Books

Bennett, J. O., Donahue, M. O., Schneider, N., & Voit, M. "The Cosmic Perspective", 8th ed., Upper Saddle River (NJ), Pearson (2017)

Carroll, B. W., & Ostlie, D. A. "An Introduction to Modern Astrophysics", 2nd ed. San Francisco, Pearson Addison Wesley (2007)

Chaisson, E., & McMillan, S. "Astronomy Today", 8th ed., Upper Saddle River (NJ), Pearson (2014)

Kaufmann, W. J. & Comins N. F. "Discovering the Universe", 10th ed., Freeman and Company, New York (2016)

Kutner, M. L. "Astronomy: A Physical Perspective", 2nd ed., Cambridge, Cambridge University Press (2003)

Morison, I. "Introduction to Astronomy and Cosmology", Hoboken (NJ), Wiley (2008)

Perryman, M. "The Exoplanet Handbook", 1st ed., Cambridge, Cambridge University Press (2011)

Seager, S. (Ed.) "Exoplanets", Tucson, University of Arizona Press (2010)

Seeds, M. A., & Backman, D. "Horizons: Exploring the Universe", 13th ed., Pacific Grove (CA), Brooks Cole (2013)

Further Readings

Unit 1

Clegg, B. "Gravity: How the Weakest Force in the Universe Shaped Our Lives" (2012)

Dolnick, E. "The Clockwork Universe: Isaac Newton, the Royal Society, and the Birth of the Modern World" (2012)

Unit 2

Napoli, D., & Balit, C. "Treasury of Greek Mythology: Classic Stories of Gods, Goddesses, Heroes & Monsters" (2011)

Rey, H. A. "The Stars: A New Way to See Them" (1976)

Tirion, W. "The Cambridge Star Atlas" (2011)

Unit 3

Hall, A. "Getting Started: Budget Astrophotography" (2014)

Legault, T. "Astrophotography" (2014)

Watzke, M. "Light: The Visible Spectrum and Beyond" (2015)

Unit 4

Berry, R. "Discover the Stars: Starwatching Using the Naked Eye, Binoculars, or a Telescope" (2010)

Penricke, S. "Astronomy with a Home Telescope: The Top 50 Celestial Bodies to Discover in the Night Sky" (2015)

Unit 5

Davis, L. "Mars: Our Future on the Red Planet" (2016)

Sobel, D. "Longitude: The True Story of a Lone Genius Who Solved the Greatest Scientific Problem of His Time" (2007)

White, E. B., & Kennedy, J. "The Moon 1968–1972" (2016)

UNIT 6

Scientific American Editors "Exoplanets: Worlds Without End" (2015)

Summers, M., & Trefil, J. "Exoplanets: Diamond Worlds, Super Earths, Pulsar Planets, and the New Search for Life beyond Our Solar System" (2017)

UNIT 7

Curtis, H. "The Scale of the Universe", Bulletin National Research Council, Vol. 2, Part 3 (1921)

Hawking, S. "A Brief History of Time" (1988)

Hubble, E. P. "The Observational Approach to Cosmology" (1937)

Johnson, G. "Miss Leavitt's Stars: The Untold Story of the Woman Who Discovered How to Measure the Universe" (2006)

UNIT 8

Begelman, M. "Gravity's Fatal Attraction: Black Holes in the Universe" (2009)

Berman, B. "The Sun's Heartbeat: And Other Stories from the Life of the Star That Powers Our Planet" (2012)

Berman, B. "Zoom: How Everything Moves: From Atoms and Galaxies to Blizzards and Bees" (2014)

Summers, M., & Trefil, J. "Exoplanets: Diamond Worlds, Super Earths, Pulsar Planets, and the New Search for Life beyond Our Solar System" (2017)

Useful Websites

https://aas.org/
https://en.wikipedia.org/wiki/Astronomical_constant
https://en.wikipedia.org/wiki/List_of_brightest_stars
https://en.wikipedia.org/wiki/List_of_nearest_bright_stars
https://en.wikipedia.org/wiki/Physical_constant
https://en.wikipedia.org/wiki/Solar_irradiance
https://en.wikipedia.org/wiki/Variable_star
http://eso.org/
https://nssdc.gsfc.nasa.gov/planetary/factsheet/
http://exoplanet.eu/
https://exoplanets.nasa.gov/
https://kepler.nasa.gov/
http://messier.seds.org/
http://physics.nist.gov/cuu/Constants/index.html
http://simbad.u-strasbg.fr/simbad/
https://solarsystem.nasa.gov/planets/
https://stardate.org/nightsky/constellations
https://www.astrosociety.org/
http://www.astronomynotes.com/telescop/chindex.htm
http://www2.ess.ucla.edu/~jewitt/kb/big_kbo.html
https://www.nasa.gov/
https://www.noao.edu